COMMERCIAL VEHICLES ARCH

C000073561

THE LEYLAND BE_____

by Graham Edge

ACKNOWLEDGMENTS

Neil Mitchell and Gordon Baron have once again come up trumps with some magnificent material from the British Commercial Vehicle Museum Trust Archives at Leyland. The team there have "beavered" away for hours in the search for interesting and informative photographs and are deserving of my gratitude. From Australia Tony Petch has sent photographs of the highest quality including some of Beaver conversions specifically for Australian operators which were unique to that vast and interesting country. Similarly Rufus Carr found Beavers in New Zealand which caused some food for thought. Arthur Murray contacted numerous transport men in New South Wales and further afield to unearth long hidden photographs. To all these mentioned and others who have become an important part of the 'Commercial Vehicles Archive Series', please accept my grateful thanks.

Photographic Credits

The majority of photographs reproduced in this book are official Leyland Motors material with the negatives in the care of the British Commercial Vehicle Museum Trust Archives. All other photographs used have been duly acknowledged in the caption. Every effort has been made to trace the original copyright holder if there was any doubt.

The Commercial Vehicles Archive Series is produced and published by Gingerfold Publications in conjunction with the British Commercial Vehicle Museum Trust Archives.

This book was first published in November 2000 by Gingerfold Publications, 8 Tothill Road, Swaffham Prior, Cambridge CB5 0JX.

Copyright

Graham Edge, Gingerfold Publications, and the British Commercial Vehicle Museum Trust Archives.

ISBN 1 902356 06 3

All rights reserved

Other titles available in this series
AEC Mandator V8
Leyland Atlantean
Leyland Comet
AEC Mammoth Major MkIII

Titles in preparation
Leyland Buses for London Transport
AEC Lorries of the 1930s
AEC Mercury

No part of this publication may be reproduced, stored in a retrieval system or transmitted in any form or by any means of electronic, mechanical, optical, photocopying, recording or otherwise without written permission from the publisher. Except for review purposes only.

Typesetting, design and printing by The Burlington Press, Foxton, Cambridge CB2 6SW.

INTRODUCTION

Certain lorry models have become symbolic of the status and reputation of their manufacturer. Without any doubt Leyland's Beavers epitomised all that was representative of a great company at its best. All too often historians and journalists dwell on the less than successful years of the late sixties and seventies at British Leyland. It is easy to understand why this is so because by then British Leyland was Britain's largest and most important car and commercial vehicle builder. But they do the old firm a great injustice and conveniently ignore the magnificent achievements gained previously.

The Leyland Beaver name survived over forty years from the first appearance of this distinguished model in 1928. Within 10 years of inception there was an almost confusing assortment of Beavers suitable for just about every task a lorry could perform. Leyland Motors had perfected the top weight four wheeler and dominated the market for this particular kind of vehicle.

After the second world war Beaver variants were never as numerous as in the pioneering years of the thirties. From 1945 for 15 years or so, wagon and trailer outfits enjoyed their heyday and Beavers were often seen in this role. These fine lorries exuded strength and reliability, becoming the archetype prime mover. As road transport modernised in the sixties so Beaver articulated lorries became increasingly popular.

Just as Beavers performed admirably at home they also served with great merit in numerous other countries. In due course Super Beavers, specifically designed for exporting, appeared and enhanced Leyland Motors' enviable overseas reputation. The expertise of designers and engineers employed by the venerable Lancashire firm was sought by European companies, and Leyland technology enabled at least two such lorry builders to achieve prominence today.

My personal recollections of Leyland Beavers are confined to observing them in my formative years. Especially, and as a schoolboy I took great interest in the seven or eight Beavers owned by E.P. Potter & Co. Ltd. of Little Lever. Firstly these were 12.B types and then they were replaced by 14.B versions. All were used for drawbar trailer work hauling poisonous chemicals around Northern England. Sometimes they carried demountable tanks of bulk acids or chrome liquor; on other occasions their loads comprised carboys, drums, or sacks. These were imposing lorries and the resonance of their engines working hard was memorable. Nor can the sound of a 0.600, in particular, ticking over ever be forgotten. Quite a few years later when Potters and their Leylands were only happy memories I was sorely tempted to purchase a former mass X-Ray Beaver as it came to the end of its lengthy working life. Happily several of these hardy stalwarts entered preservation and their presence grace rallies and similar events around the country.

The history of Leyland Beavers is quite complex because of the longevity of most types in production. This study covers all main versions and their development. The majority of photographs selected are from the vast quantity held by the British Commercial Vehicle Museum Trust Archives and are prime former Leyland Motors material.

Graham Edge
Swaffham Prior, July 2000

SECTION 1
Early Beaver Lorries

Leyland Motors Ltd. was incorporated in 1907 although the enterprise had existed and expanded under other trading names previously. Its origins being in a small smithy which soon encompassed general engineering before starting to build steam wagons. Early in the twentieth century the first internal combustion engined lorries were made at Leyland. Soon these were developed to become solid and reliable. During the Great War of 1914–18 Leyland Motors became a major supplier of lorries to military procurement departments.

After rapid expansion during the first world war Leyland Motors then experienced a very difficult period in the early 1920s. Demand for commercial vehicles of all kinds was poor and the company suffered serious financial losses. However, Leyland Motors survived and by 1923 sales were improving sufficiently for recovery to commence. By then Leyland products had already become recognised for high quality designs and engineering. Indeed the company was one of Britain's largest commercial vehicles manufacturers with many models achieving market leading status. Leyland Motors made most of its components at production facilities in and around the Lancashire town from which the company took its name. They also had a sizeable factory at Kingston-upon-Thames.

By the year of 1926 Leyland Motors was advancing once again due to the success of its 'L' bus range, introduced in 1925 and designed by the highly regarded J.G. Rackham. Within three years over 4,000 'L' types were sold with the Lion being most successful and popular. There cannot be any doubt that J.G. Rackham was a brilliant engineer and designer, but the then General Manager of Leyland Motors, A.A. Liardet, also had a profound influence on the company's fortunes. He established a programme of rationalisation and re-organisation which was tremendously beneficial. Many designers would have been content to develop designs as popular as 'L' types, but Mr Rackham was not a man to rest on his laurels. By late 1927 yet another acclaimed passenger chassis range was in being:– the famous 'T' types which included the Titan TD1 double-decker. This was truly a revolutionary model, the first bus of its kind to carry a low height body.

With the popularity of 'L' and 'T' types assured messers Liardet and Rackham next turned their attention to Leyland lorry models. Existing types were basically versions of wartime designs from ten years before, and by then were rather dated. Transport legislation for lorries had remained conservative and most types and sizes were still restricted to a maximum speed limit of 20 mph.

The successful Leyland passenger ranges were planned around new petrol engine and gearbox designs which also were perfectly suitable for lorries. Such was the rate of progress then that new engine types apparently were announced almost annually. There were 4 and 6 cylinder versions with various cylinder dimensions but usually a common piston stroke length. By 1928 many of these Leyland engines were of overhead camshaft layout (ohc), but the camshaft still opened the valves through rockers rather than by direct action. Nevertheless a reliable, successful design which contributed enormously to Leyland's growing reputation.

At the Olympia Motor Show of 1928 Leyland Motors proudly displayed its first new, modern lorries for over a decade. Fresh model names were introduced with the Leyland Beaver four wheeler making its debut. Even then in the late twenties some lorries still had solid tyres and partially open cabs. Not so with the Beaver however, it was shod on pneumatics with a fully enclosed driver's structure. For then it was an advanced lorry with attractive styling complemented with an up to date 4 cylinder petrol engine and 4 speed constant mesh gearbox. The Beaver was rated for a carrying capacity of 3½ tons in normal, control, bonneted, configuration.

Along with other Leylands the Beaver made a huge impact at Olympia and these lorries attracted numerous orders for the company. With a proven, reliable, and smooth power unit the Beaver had a massive advantage over many of its rivals. From the outset export versions were available and excellent sales were gained overseas with other contemporary Leylands also proving very popular. Leyland Motors was becoming a highly respected company worldwide and by 1932 had representation in many countries.

Immediately on its launch in 1928 the Leyland Beaver was recognised as a top quality lorry and this first model spawned several versions over the following years. If lorry design had stagnated in the previous decade it was completely different in the thirties with continuous development and constant advancement. But before then there was another grave world-wide manufacturing depression to overcome. Even though Leyland Motors had enjoyed several boom years with its successful bus and, latterly, lorry models, it was not immune from recession and profitability suffered badly. Additional share capital had to be sought and the company was re-organised and re-named Leyland Motors (1933) Ltd.

The opening years of the thirties were memorable. Trading conditions were harsh at home and abroad. But it was a time of innovation and the first diesel (or oil) engines for automotive purposes were introduced. In addition the road transport industry was subjected to radical legislative changes. New statutes were to influence lorry design and operations for the next twenty years. For instance, much greater emphasis had to be placed on achieving lower unladen weights to maximise payloads within gross vehicle weights. Consequently chassis designs had to be re-worked to lighten them. Administrative matters were also overhauled with the introduction of Carriers Licensing:– the well remembered 'A' and 'B' Licences for hire and reward hauliers, and 'C' Licences for own account operators.

New Construction and Use regulations sounded the death knell for steam wagons which therefore encouraged demand for internal combustion engined lorries. Although regulatory laws were partly designed to protect the freight interests of railway companies, nevertheless the thirties witnessed a massive growth of road transport. Much of this business was won from railways which simply could not compete with the improving efficiency of road haulage. Even though lorries were subjected to low speed limits they could deliver goods from door to door much quicker than rail freight. As overnight trunking services became established merchandise loaded in Lancashire one day would be delivered in London next morning, a service which railways simply were incapable of matching.

By the end of 1933 Leyland Motors had recognised the legislative implications. They offered a revised Beaver range which was extended in the following months. Prospective customers were offered a comprehensive choice of models: 4 and 6 cylinder petrol or diesel engines; normal (bonneted) or forward control layout. This 1993/34 line-up was listed for the reminder of the decade with specification improvements at regular intervals. The main types available, along with their alternative propulsion units are dealt with individually. Appendix A provides further details of the entire Beaver lorry family from inception in 1928 until cessation of production over forty years later.

Beaver 4:– TC Normal Control, TSC Forward Control

Leyland Beaver 4 lorries were rated for 6 to 8 tons payload within a maximum gross vehicle weight of 12 tons. They were powered by 4 cylinder engines, either petrol or diesel which produced approximately 70 bhp, (adjusted to BS AU141: 1967 criteria). They were available with a choice of two load carrying wheelbases in either normal or forward control chassis. There was also a tractive unit, again this was listed in both formats. Usually lorries for tipping duties had deeper main frame side rails made from slightly thicker steel. From 1937 a heaver and stronger chassis was provided for all applications and these models were given revised type designations with an 'A' suffix after the number.

Whichever engine was used a generous 16¼ inch single plate clutch, with clutch stop, was fitted. A 4 speed constant mesh gearbox, unit mounted with the engine, was provided and this was a "silent third" design. In other words the third ratio consisted of helically cut gears rather than straight cut ones. Helical gears are always quieter and it was a fashion of the 1930s to promote "silent third" gearboxes. Fourth speed was direct drive to a overhead worm and wheel rear axle with a ratio of 7.33:1. An optional diff of 6.5:1 became available in 1939. From about 1937 a 2 speed auxiliary gearbox could be specified to special order. This was a 'step-down' or reduction unit with a ratio of 1.54:1, low speed, aimed mainly for tractive unit purchasers.

Beaver 4s had Marles cam and roller steering boxes. By the standards of the time they also had good brakes which were superior to those of many competing lorries. Front brake shoes were 3 inches wide with those at the rear 6 inches. Brake drums were 17 inches diameter. The foot pedal operated a master vacuum servo actuator which transmitted the driver's efforts to the back brakes by a mechanical cross shaft. Individual slave vacuum servos were used on the front wheels. The parking brake operated on the rear wheels and was a push-on lever in the driver's cab. Usual tyres were 36 x 8 inches fitted onto ten stud wheels. Fuel was drawn from the main tank to either the carburettor or diesel injection pump by an autovac.

Beaver 6:– TC Normal Control, TSC Forward Control

The Leyland Beaver 6 range was a heavier duty line-up than contemporary Beaver 4s, intended primarily for trailer work. They could be either drawbar or articulated semi-trailer types. Solo gross vehicle weight rating was 12 tons, with up to 19 tons gross train weight for combinations. Again there were three wheelbase options in either normal or forward control layout. Heavier and stronger chassis were supplied from 1937. Propulsion was from 6 cylinder petrol or diesel engines and initially the petrol unit was the more powerful, producing approximately 110 bhp. The diesel unit was some 17 bhp less powerful at first, but was re-tuned later to give similar power to the petrol engine.

Beaver 6s had slightly heavier chassis than their stablemates with frames being deeper and thicker. Originally (and for most Beaver 6 models) the transmission was identical to that of Beaver 4s, but the 2 speed auxiliary gearbox was available from 1936. For tractive units a heavy duty 4 speed constant mesh unit was used. This was replaced in 1938 by a 5 speed constant mesh unit. Still a direct drive design it has a low crawler ratio of 7.75:1 and closer ratios in the intermediate gears. This gearbox was offered as an alternative in other Beaver 6 chassis. Because Beaver 6 tractors were suitable for heavy work an optional back axle was listed from the outset. It was a double reduction spiral bevel arrangement with a standard ratio of 7.15:1. Lower or higher gearing options could be ordered by the customer.

Steering, brakes, tyres, wheels, and fuel systems were identical to those of Beaver 4s. From 1934 all Beavers received deeper radiators with increased coolant capacities. These lorries were modern with an imposing profile. Most purchasers opted for Leyland's own cab, but coachbuilt structures from other makers could be fitted if that was the customer's choice.

Petrol Engines

By the early thirties Leyland Motors possessed an excellent range of petrol engines. They were noted for smoothness and refinement, evolving from designs introduced in the mid-twenties. Leyland produced many components in-house including castings from the company's foundry. One important feature of all pre-second world war Leyland engines was the extensive use of aluminium alloy for low weight. Leyland's petrol engine family was comprehensive and included 4 and 6 cylinder versions which shared common bore and stroke dimensions along with other design features. The engines now described were standard units listed for Beaver lorries. However, such was the choice at this time others could have been fitted at the insistence of certain operators. To give meaningful comparisons and reference points for power outputs the figures quoted are to the now obsolete British Standards AU141:1967 ratings.

Beaver petrol engines were overhead camshaft layout with the camshaft alongside the valve stems. Rocker levers operated the valves but the need for push rods was obviated. The cylinder block was a monoblock ferrous casting and the crankcase was cast from aluminium alloy. The pistons were of similar metal. Single piece cylinder heads were used and could be either cast from iron or alloy, depending upon the engine version and application. Chain drive was used for the camshaft and auxiliaries. Four cylinder units had 5 bearing crankshafts and six cylinder engines had 7 main bearings. All those years ago the old RAC horse power rating was an important measure of engine size and even dictated the taxation class of vehicles. Appendix B provides details of all Leyland Beaver engines for the duration of production, but the salient differences between 4 and 6 cylinder types are as follows.

4 Cylinder Engine

The bore of 4.56 inches and stroke of 5.5 inches gave a swept volume of 360 cubic inches (or 5.893 litres). With an RAC rating of 33.3 horse power. This was equivalent to approximately 70 bhp at 2,200 rpm. Torque output was 250 lb ft. at 1,100 rpm. Ignition was by a Simms magneto and a 42 mm Zenith carburettor was fitted; this had an automatic choke. A starting handle was the usual method of firing-up, but a 12 volts electric starter motor was optional and an oil cooler was fitted as standard.

6 Cylinder Engine

With a bore of 4.25 inches and stroke of 5.5 inches the swept capacity was 468 cubic inches (or 7.670 litres). The RAC rating was 43.5 horse power which equates to 100 bhp at 2,200 rpm. Maximum torque was 320 lb ft. at 1,000 rpm. A CAV-Bosch magneto provided the means of ignition and a 46 mm Solex carburettor was standard. Again this had automatic choke for cold starting. A 12 volts electric starter motor was available to order and as usual with all engines containing a high percentage of aluminium, an oil cooler was fitted.

Both 4 and 6 cylinder engines had rather "peaky" torque curves with pulling characteristics optimised within rather a limited band of engine revolutions.

It is also worth mentioning Leyland's Mark III 6 cylinder petrol unit, available in 1937. Conceivably this engine could have powered some Beavers. It was slightly larger at 482 cubic inches (or 7.900 litres) and was a conventional overhead valve, push-rod layout which had two alloy cylinder heads each covering three cylinders. Power output was similar to that of the ohc 6 cylinder unit.

Diesel Engines

Until the early thirties all British made internal combustion engined road going vehicles were fuelled by petrol. While Dr Rudolf Diesel's compression ignition theory had been adapted for marine and industrial purposes from the early years of the last century, constructing satisfactory diesel fuel automotive units proved difficult. The main problems centred around finding a suitable fuel injection pump design for a smaller engine revving much faster than heavy marine and industrial units. The German company Bosch pioneered fuel injection equipment and by the mid-twenties some European lorry builders, notably Saurer, had diesel (or oil as they were commonly known) engines in production. Even so, many early units tended to be rough running, smelly, and of dubious reliability.

The economics of diesel propulsion were attractive because this fuel was approximately one third the cost of petrol. Incredible as it might seem today, petrol was then the equivalent of 6 pence a gallon and diesel just 2 pence. Added to this fuel economy was also better. Both of Britain's foremost heavy commercial vehicle builders, Leyland Motors (1927) and AEC (1928) had produced their own diesel engine designs by the end of the twenties. But these were prototypes for their own evaluation purposes, unsuitable for offering to customers. The acclaimed engine manufacturer Gardner of Patricroft showed the way forward and in 1931 introduced their famous LW automotive diesel engines.

By 1933 Leyland Motors had acceptable diesels available which could be sold as feasible alternatives to petrol units. Generally they were remarkably similar in design and layout to Leyland petrol types and were available with 4 or 6 cylinders. Obviously there were significant differences dictated by fuel ignition requirements. But Leyland was unique in being able to produce a diesel engine no larger in its overall dimensions and therefore interchangeable with equivalent Leyland 4 or 6 cylinder petrols. This made re-engining a petrol chassis with a more economical disel unit an easy proposition. Remember that in those days premium lorries such as Leylands were built to last and an operator would expect up to 15 years service. So, retro-fitting with diesel engines was viable and became very common. To specify a diesel engine in a new chassis cost some £175 for a 4 cylinder and £275 for a 6 cylinder. Diesel engines were more costly to manufacture but service life was longer and with better economy from cheaper fuel, additional purchase price was more than recovered during the vehicle's lifetime.

These Leyland diesel engine designs of 1933 remained mainly unaltered in their basic formats until the outbreak of war 6 years later. Within a few months indirect injection was abandoned in favour of direct injection and power outputs were increased during the decade. They were strongly based on the successful Leyland petrol range with identical ohc configuration. Crankshafts, main bearings, and crankpins were bigger to better cope with specific stresses imposed by the compression ignition cycle. Crankshafts were also fitted with dampers. Initially varying bore dimensions were offered, all with a common piston stroke length. For Beavers, 4 and 6 cylinder units had common bores of 4.5 inches and strokes of 5.5 inches. The 4 cylinder engine was 350 cubic inches (5.700 litres) whilst the 6 cylinder was 525 cubic inches (8.600 litres). The former engine's crankshaft was carried in 5 main bearings with of course 7 mains for that of the latter. Weight increases were kept to a minimum over petrol units with the diesels being just 1 cwt and $1^1/_2$ cwts heavier respectively.

Initially Leyland's new diesels were indirect injection types with pre-chamber cylinder heads. However, within months and by 1934 this concept was abandoned. Like competing Gardner engines Leyland Motors then employed direct fuel injection. Bosch fuel pump designs were licensed to C.A. Vandervell in Great Britian and Leyland fitted a CAV-Bosch injection pump with fixed timing. The 4 cylinder unit had a maximum governed speed of 1,900 rpm and produced 68 bhp at 1,800 rpm with 205 lb ft of torque at 1,300 rpm. Its bigger sister was governed slightly slower at 1,850 rpm and at first gave 93 bhp at these revolutions. Torque was 305 lb ft at 1,300 rpm. By 1939 this 6 cylinder engine had been re-tuned and up-rated to an output of 106 bhp at 1,900 rpm. Idling speeds were 400 rpm and 350 rpm respectively for 4 and 6 cylinder types.

Leyland used its own design of combustion chamber in the piston crown. It was a simple cup, or pot, into which fuel was injected by a Leyland made single hole sprayer, as injectors were called then. Until diesel fuel quality improved and filtration methods became reliable multi-hole sprayers tended to block easily resulting in reduced power, rough running, and smokey exhausts. These early Leyland diesels were quite good starters and pre-heating glowplugs were unnecessary. Electric starter motors of 24 volts were fitted as standard and a de-compression lever was provided to facilitate easier starting. A rotary exhauster and chassis mounted vacuum tank were fitted for the braking system servo units.

Whilst both diesel engines had lower torque outputs than the contemporary Leyland petrol units, they had flatter torque curves which meant they would pull better over wider engine revolutions range. In particular the 6 cylinder diesel was a good 'slogger' at medium revs.

With reliable diesel engines available at last sales of Beavers and indeed all Leylands were excellent throughout the thirties. The company prospered with exports becoming a very important segment of the business. By 1937 almost two fifths of output was destined for overseas markets.

Without any doubt the Leyland Beaver range pre-war was comprehensive with a large choice of petrol or diesel powered lorries available. It would have been a peculiar customer indeed who could not buy a Beaver suitable for his needs.

What is believed to be the first Leyland Beaver, introduced in 1928, was sold to G. & P. Barrie Ltd., Scottish mineral water manufacturers. Even at that date some rival's lorries had partially open cabs and solid tyres. Not so the Beaver, as a completely new model it was thoroughly up to date and was robustly constructed to handle a nominated payload of 70 cwts.

A typical mixed load of provisions (groceries) of the time on MacSymon's Beaver. The barrel almost certainly contained butter and there was no refrigeration widely available then in the early thirties. Liverpool's Custom House was bombed during the Second World War and demolished, and the location shown here is now the site of Central Fire Station.

This is a fascinating picture of a 4 cylinder Beaver painted in the livery of Redfearn Brothers Ltd. of Barnsley, glass container makers. The lorry's load is definitely not glass bottles. and appears to be bales of wool. The driver seems to have made a complete mess of sheeting the load. Note the Shire horse's hooves visible underneath the Leyland.

For the first five years of Beaver production only the 4 cylinder normal control model was made, but it was very popular in many diverse roles. This is road making the hard way, before the use of Barber Greene spreaders for tarmacadam.

Leyland Motors quickly established good export sales with its Beavers with many of the old British Empire countries proving to be loyal customers. These long wheelbase tankers were based in Toronto Canada, in service with the McColl Frontenac Oil Company.

Australian companies bought thousands of Leyland lorries through the years, and The Yungburra Sawmills of North Queensland placed a couple of six cylinder diesel engine powered Beaver 6TCs in service in the mid-thirties. Presumably a cab was thought unnecessary in sub-tropical Australia.

By the time these Beaver TSC lorries entered service in the mid-thirties, carriers J. Nall had been in existence for some 90 years, since 1845. With a head office in Manchester, they had several depots in Lancashire, and the Beavers photographed here were based at Bolton. All are loaded with either raw or semi-finished cotton and were serving the busy cotton spinning mills, which were numerous in Bolton then.

Cunningham's Brewery of Warrington operated this unusual Beaver brewer's dray and trailer. Seen here delivering to the now demolished Bankes Arms in Wallgate, Wigan. The fixed roof must have made it difficult for the crew to load and unload double-stacked barrels, although the absence of traffic in the thirties meant it was possible to unload into the middle of the road!

There is plenty of interest in this picture, probably taken at the Battersea Flour Mills of Hovis. When these Beaver 6 TSCs entered service in the thirties all flour was delivered in sacks that weighed 2.5 hundredweights each and bulk flour delivery by tanker was still twenty years in the future. AXU 314 has a drawbar trailer attached, and note the van bodied steam waggon.

An export Beaver 6 TSC tractive unit sold in South Africa to W.H.J. Heins. By British market standards it has a long wheelbase and a short tandem axle dropside semi-trailer. The absence of mudguards on the rear axles is interesting, and reflects local practice.

Although this is a carefully posed shot for the photographer it is still full of character, depicting a typical quarry scene of its era. Midlothian County Council operated several Leylands in its fleet and there is another contractor's Leyland in the background with a couple of early Bedfords on the upper level.

Since opening the first Mersey Tunnel the authorities have operated their own recovery fleet to facilitate the rapid removal of broken down vehicles. Their Beaver breakdown lorry was supplied in July 1934 with a Leyland built body and Morris crane. This vehicle is believed to have survived into preservation, although it has not been seen in recent times. Note the obviously quite new "Belisha Beacon", but the absence of familiar Zebra crossing stripes.

Compare this photograph of loading less heavy sacks of cement with sack trucks to that of loading flour. The load for each of these Beavers was 8 tons, and unusually the sacks were being stacked on end, rather than flat as is usual with cement. Would loading have continued in the rain? The photo dates from 26th October 1937.

It was a wet day in Bispham when this abnormal load movement took place using a Beaver and extended drawbar trailer. The steel column was 50 feet long and was being removed from the Hoo Hill Brickworks. It appears to be quite precarious, with a minimum of restraining chains to secure it. Obviously heavy, note the bending of the trailer because of the weight imposed on it.

On 12th May 1938 J. Saxton's Beaver cattle float was seen leaving Huddersfield Borough slaughterhouse. What an impressive entrance archway with sculptures of livestock incorporated into it.

Another carefully posed scene for the benefit of Leyland Motors' photographer, but providing an insight into working conditions of 1937 for drivers. They were expected to clean their lorries and probably perform routine, basic, maintenance and servicing. These Marsh & Baxter men were quite an elite band, with better than average conditions (compare this spacious warehouse with the cramped Hovis flour mill). The "Refrigeration Van" would be carrying 'dry ice', i.e. frozen carbon dioxide.

Almost from inception to the articulated revolution of the sixties, various Beaver models have been ideal prime movers for trailer work. In the thirties this potential was realised and Beavers were the preferred choices of many hauliers whose lorries pulled trailers. This well laden Beaver 6 TSC tackles a steady, rising incline in rural Lancashire.

SECTION 2
Post-War Progress

To recap the status of Leyland Motors and its Beaver range in 1939. The company had prospered to become Britain's largest maker of top quality commercial vehicles. Along with its fierce competitor AEC it vied for the title of most influential manufacturer and designer. Leyland Beaver lorries were reliable and durable but importantly, with unladen weights low enough to qualify for the under 5 tons unladen weight taxation class. Not only was Leyland Motors' prestige high at home it was successful world-wide. Already it had become an accomplished promotor of its products through ambitious advertising, and marketing activities.

When the country went to war in September 1939 shortly afterwards production of most lorries for civilian customers was halted. Leyland's factories were turned over to a multitude of tasks for the war effort. Vehicles for the armed forces, notably Leyland Retrievers and Hippos continued to be made in Lancashire. At Kingston-upon-Thames lightweight Lynx models remained in production for most of the duration. Many of these chassis again were destined for the War Office, but a reasonable quantity were released for essential non-military needs.

As the conflict progressed Leyland made engines for several applications including fire pump power units, and generators. The company soon became an important and vital battle tank manufacturer and this involvement had a direct influence on engine development. Initially the overhead camshaft Beaver 8.6 litre diesel was modified for twin installation in Matilda tanks. However, what was a proven and reliable lorry and bus unit was found wanting in heavy military equipment. With a single, long, flat cylinder head engine dependability was poor when subjected to widely varying operational loadings with intense cycles. Also its high aluminium alloy content caused difficulties as supplies of this metal were urgently needed for aeroplane manufacture.

In mid-1941 a new mainly ferrous metal engine was fitted into some tanks and again this was a twin installation. Somewhat smaller than its predecessor at just over 7.1 litres capacity it produced 105 bhp at 2,000 rpm. This design was modified for road-going usage a couple of years later when a slightly larger bore version was fitted into Leyland Hippos. Now of 7.4 litres capacity it produced 100 bhp at 1,800 rpm and was a conventional overhead valve, push rod layout with two cylinder heads, one for each set of three cylinders. There was another important motive, power development from that period which became one of the great Leyland engines of all time. The renowned 9.8 litre "600" which could be specified to run as either a petrol or diesel fuel unit and is described in more detail shortly. With all Leyland's considerable manufacturing capacity at full stretch during the war years the workforce almost doubled from 6,500 employees to 12,000, of which some 25 per cent were female.

Even with a return to peace in 1945 there could be no let-up in demands for Leyland products. Not only were customers at home requiring large numbers of new buses and lorries to replace elderly pre-war vehicles, there was also a government inspired export drive. This was a means of creating badly needed foreign exchange for the country to repay its accumulated wartime debts. Raw materials, rationing and shortages caused immense difficulties which were studied in greater details in another Commercial Vehicles Archive Series volume, – 'Leyland Comet'.

A completely new vehicle range was planned by Leyland Motors for their post-war catalogue. But until re-organisation of their factories for peacetime working conditions could be completed some types were mixtures of old and new features. In particular the first Beavers sold in 1945 and 1946 were interim models. They still retained many pre-war characteristics of chassis and cab design but were powered by the recent 7.4 litre engine. Versions of Beavers available from that time onwards were never as copious as during the thirties and for home markets normal control models were discontinued.

Later in the year of 1946 thoroughly modernised 12 tons gross weight Beaver models were introduced. Designated 12.B.1 or 2 or 3 there were just three wheelbases available:– a long wheelbase rigid ideally suited to drawbar trailer work; a sturdy tipper; and tractive unit rated at 19 tons gross train weight. The chassis with minimum 10 inches deep frames retained its pre-war reputation of ruggedness and strength, and typically the longest Beaver with a flat platform body tared off at about 5 tons, allowing only 7 tons of legal payload. These Beavers were over engineered for such modest loads and were capable of carrying more. They were perfect for trailer duties. Legally the articulated tractive unit could carry about 12 tons payload, depending upon its semi-trailer.

Not only was this new Leyland strongly constructed it had a powerful engine by the standards of the mid-forties. The novel 9.8 litres design known as the 0.600 (0 = oil engine) produced 125 bhp at 1,800 rpm and 410 lb ft of torque at 900 rpm. The cylinder bore was 4.8 inches (122 mm) and stroke retained its pre-war dimension of 5.5 inches (139.7 mm) giving a swept volume of 597 cubic inches. Both crankcase and cylinder block were ferrous castings and the dry liners were pre-finished and easily replaceable with the engine in situ. Two cast iron cylinder heads, each covering three cylinders were fitted, with pushrod activated overhead valves. The camshaft and auxiliaries were gear driven. The crankshaft was turned from an alloy steel forging and its large crankpin and main bearing journals were nitrided. Balance weights and a vibration damper were incorporated and with 7 main bearings the total surface area was generous.

Usually a Simms fuel injection pump was used but CAV pumps could also be fitted. This type was the usual fitment on export engines. Leyland multi-hole injectors sprayed diesel into a toroidal cavity combustion chamber in the piston crown. The single dry-plate clutch, with clutch stop, was still 16¼ inches diameter and had four stage adjustment and a flexible centre for smoothness. Unit mounted with the engine was a new 5 speed direct drive constant mesh gearbox and the up-dated back axle remained faithful to traditional worm and wheel drive. Standard ratio, which was 6.5:1 gave a top speed of about 33 mph and initially three lower ratios were available (later reduced to two optional ratios). Marles cam and double roller steering was retained and usually the tyres were 36 x 8 inches covers on ten stud wheels. Later, modernised 9.00 x 20 tyres became standard.

Wedge operated brake shoes were used with drums of 16¼ inches diameter. Brake liners were 3 inches wide on the front and 5 inches wide on the rear. The total braking area was 509 square inches. Footbrakes were servo assisted with hydraulic cylinders on each wheel. A pull-on type of handbrake operated on the back wheels only.

A brand new all metal Leyland built cab was fitted to all Leyland heavy duty lorries and while rather angular it was not unattractive with the radiator behind a mesh grille. These earliest cabs were very plain but as materials rationing eased in later years bright metal embellishments were added.

In 1951 Leyland Motors produced a bigger engine of 677 cubic inches, or 11.1 litres. Known as the 0.680 it had a bore of 5 inches (127 mm) and stroke of 5.75 inches (146 mm). With an output of 150 bhp at 2,000 rpm and 450 lb ft of torque at 1,100 rpm, it shared many design characteristics with the Leyland 0.600 unit. This engine was prompted by demands for more power from overseas customers but it was made available for home market multi-axle Hippos and Octupuses. By 1953 it could also be ordered as an option, at extra cost, by Beaver operators, especially if a drawbar trailer was being pulled.

Also later in 1953 Beavers were equipped with better brakes, courtesy of a totally new air braking system. To distinguish between earlier models and the latest revised numerical suffixes were given; ie 12.B/8 or 9 or 10. The rationale for switching to air pressure braking by heavy lorry makers is explained in 'AEC Mammoth Major Mk.III', another volume in the Commercial Vehicles Archive Series. Leyland chose a cam-operated internally expanding method of application with actuation from individual chambers for each wheel. Drum diameters were increased to 16¾ inches with wider linings front and rear. These were 4 inches and 6 inches wide respectively. The total braking area was increased to 575 square inches. Again the parking brake was operative only on the rear wheels but the lever became a multiple pull ratchet type for better effectiveness. By then rigid Beavers could also have an auxiliary gearbox within their transmissions. This was a 2 speed unit with ratios of 1.328:1 and 1:1 which could be arranged as either reduction or overdrive depending upon the work the lorry was doing. There was also a slight revision to the rear axle specification and a ratio of 7.31:1 became standard. Optional higher or lower gearing was available.

By 1954 Leyland Motors was producing an improved version of its driver's structure. Still of all-metal build it featured an insulated double skin. Externally it presented a softer image with bright metal louvres in place of the mesh grille. There was now a bumper bar and internally the driver and mate luxuriated in winter with a heater and demister. Twelve tons Beavers equipped with the modernised cab were designated by an 'A' suffix after their chassis type nomenclature.

In the year of 1955 an overdue revision of Construction and Use regulations permitted an increase in gross weights to 14 tons solo, 24 tons for an articulated outfit, and 32 tons for rigid multi-axle wagon and drawbar trailer. Remember, the last weight increases had taken place back in 1933 and during the 20 years which had passed lorries had greatly improved. Leyland Beavers required very few, if any, modifications to comply with the increased weights and the types became 14.B 8 or 9 or 10. All 14 tons Beavers had Leyland's new style cab with a choice of either 0.600 engine (standard) or 0.680 unit (optional). Larger 10.00 x 20 inches tyres were fitted for heavier gross weights. Apart from continuous improvements in component design and production, basic Beaver specifications remained unaltered until the introduction of Power-Plus models in 1961.

By the end of production of 14 tons Beavers these lorries had set standards which other builders of similar heavyweight lorries struggled to match. They were exceptionally tough, robust, and unbreakable with prodigious longevity. Leyland 0.600 engines in particular achieved unprecedented mileages before needing attention or any repairs. Operators with good drivers recorded up to half a million service miles which for most lorries of the forties and fifties was unheard of. At home they were market leaders in their class and became the definitive four wheeler and trailer combination. Beaver tractive units were also very popular in those days when articulation was in the minority for most general hauliers. Export versions of Beavers enhanced Leyland's reputation for quality and reliability abroad and helped the company to maintain many years of prosperity and profitability in the fifties.

The nearside headlamp mask determines the date of this photograph as during the Second World War. On a snowy morning the driver and mate of Tear Bros' Beaver 6 TC tractive unit prepare to move their heavy bulldozer to another wartime airfield construction site. Indeed a rare lorry photograph from those years.

This is a nice line-up of part of Leyland Motors' own works transport fleet laden with export orders in June 1946. The three four wheelers laden with tyres and packing cases were, in fact, pre-war Beavers cabbed with Leyland's own streamlined structure. At the rear are a six wheeler Hippo and an Interim Beaver.

Interim Beavers were decidedly old fashioned in appearance, even by the standards of 1945. Nevertheless they exuded an aura of ruggedness and reliability. Evans of Widnes' lorry was loading sawn timber when about eighteen months old in 1947. Note the attire of its driver; still wearing his army battledress tunic, a common sight amongst working men in the years immediately following the Second World War.

A lovely photograph of smart looking Interim Beaver, taken in June 1951. Although the lorry was registered in Hertfordshire, it was based at BRS Carlisle and would have been acquired through compulsory nationalisation of road haulage. The Massey Harris combine harvester was from that company's Kilmarnock factory, no doubt a return load for the Leyland.

Jarvis Robinson Transport (JRT) was one of those transport companies peculiar to major seaports such as Liverpool, where they were based. The vast majority of their traffic originated from the docks and they used lorries uniquely adapted for their needs. This drawbar tractor Interim Beaver was typical of their operations and they filled lorries' tyres with a solution of water and calcium chloride to assist traction on the cobbled streets of Liverpool where wheel spin was a problem, as was lack of adhesion when braking in wet conditions.

There are 25,000 loaves of bread on this 12.B Beaver and trailer, or put another way, this 14 tons of flour from the CWS Flour Mills at Silvertown would make that quantity. These hessian sacks of flour (5 shillings deposit per sack) always rode well and a couple of strands of rope across the rear was all that was needed on a fine day with no threat of rain and need for sheeting. Pure nostalgia!

It has been suggested that this is a typical example of Scottish thrift, with this early 12.B Beaver tipper being used as a prime mover for the Ruston Bucyrus 10RB shovel. Unusually the lorry is not carrying a load as well!

Another quite early 12.B Beaver well laden with sheep in its three deck livestock body. The open top deck had fold down sides for when the lorry was used for carrying pigs or cattle. In 1951 this Leyland and four other lorries were acquired from Kelso & Sons by the Road haulage Executive, (BRS).

Nowadays people are used to buying their potatoes washed and cleaned in pre-packed polythene bags from supermarkets. This is how it used to be, in 1953 in fact. The spuds were packed into 1 cwt hessian sacks in the field from the riddle, and manually loaded onto the lorry. There is 6 tons already on the trailer, and at least 8 tons will go onto the 12.B Beaver. Destination: one of the large town or city wholesale markets, trunked through the night, and they will be on sale at a High Street greengrocer the following morning. "Five pounds of spuds please".

Lumley-Saville of Stratford-on-Avon placed a couple of air-braked 12.B Beaver tractive units in service late in 1954, and they were photographed the following January. They were being used for low-loader duties with a couple of Ruston Bucyrus 19RB excavators. The logo on the cabs suggests this firm were agents for International Harvester equipment as well, probably for the tracked crawler tractors made by IH.

Here is a long load with a difference, the load being special roof trusses being transported in February 1956 by a 12.B Beaver from BRS Dundee, (Central Group). The two-wheel dolly appears to be unbraked, so the lorry would have to bear the total braking effort. This is typical of the jobs tough and rugged Beavers were given.

A travel stained 12.B Beaver with a load of telegraph poles and the varying height of the lorry's bolsters suggests that this was regular work for the Leyland, and its owners, B.B. & H. Ltd.

Difficult as it might be to imagine now, British cars such as Austin A30s (top deck) and A40 Somersets (bottom deck) were sold in Canada in the fifties. North Coast Transport, based in British Columbia, operated this left hand control Beaver tractive unit as a car transporter, complete with sleeper box for the driver. The outer rear axle tyre on the Leyland is bald, but the others in view seem to have good tread depth. Maybe there had been a puncture and a 'get-me-home' spare wheel had been fitted.

The fitting of a "modernised" cab in 1954 altered the appearance of the Beaver range, and along with air brakes as standard an already excellent lorry was further improved. Alexander Scott of Paisley purchased twenty Beaver 14.B/10 tractive units and BTC 4-in-line trailers to replace their elderly Scammell artic and rigid eights. By 1962 Scotts were running sixty-five lorries and the business was acquired in that year by Smiths of Maddiston

Beavers could be bought as chassis and running gear only for special bodies, and Bonallack built the cab and body for this 14.B model. It is not a particularly attractive design, but the single piece windscreen and curved side windows were futuristic for the mid-fifties. Smiths made instruments and components for the automotive industry.

Fellside Transport Ltd. was formed to specialise in livestock transport after R.Kelso & Sons was nation-alised. Compare this body design with the earlier Kelso's Beaver, as the trailer has a lower third deck (or basement?). York Trailers built the trailer chassis and running gear at their Corby factory, but the body was made in-house. This livestock float could carry up to 350 lambs.

Passing along English Street, Dumfries, near to its home base of Cresswell Mill, Wyllies' 14.B Beaver is carrying a load of hay bales and assorted sacks of animal feeds for local deliveries. This would be a fairly easy day for the Leyland; its other regular duties were running to Glasgow docks and national com-pounders' mills around that port with its drawbar trailer for a heavy load back to Dumfries.

By the late fifties maximum weight articulated lorries were gaining popularity. Certain specialist vehicles were being designed and Bristol Engineering Company made this dedicated hydrogen gas carrier, based on a BTC 4-in-line trailer and Beaver tractive unit. The vehicle would be almost up to its maximum of 24 tons gross empty because of the weight of the pressurised cylinders; a full consignment of hydrogen weighing only 3 cwts.

Frig-Mobile of Australia Ltd. had several depots and this Beaver artic was typical of their operations. This Beaver had a longer wheelbase than home market tractive units and had twin 55 gallons capacity tanks for long distance duties. The trailer bodywork was made from aluminium alloy with 6" to 8" of insulation to maintain a temperature of minus 10 degrees Fahrenheit. The rear bogie had air suspension utilising air bags as per modern practice. (Photo: Young & Richardson, from Tony Petch).

Beavers destined for Australia were virtually identical in appearance to home market models. The mechanical specifications differed in certain aspects, but again, they were basically the same. This Beaver artic was about to enter service with John Darling & Son Ltd., Flour Millers of Adelaide. Note the roof mounted marker lights. (Photo: Ringwood Studios, from Tony Petch)

New Zealand also had its fair share of Beavers and this one belonging to Transport (Nelson) Ltd. from South Island was still going strong when this photograph was taken in 1975. By then the lorry was showing signs of age, but it was at least fifteen years old. It was powered by an 0.680 engine. (Photo: Rufus Carr)

There is quite a long trailer behind this Beaver, and it is certainly a "long-pin" type with that much expanse ahead of the fifth wheel. A.W. & T.H. Ringwood of Gepps Cross, South Australia operated the Leyland. At the time of this photo in 1962 the driver was taking a short break at the Buronga Service Station (Leyland, Albion, Scammell agents). Buronga is near Mildura in north west Victoria and is a major rural service area, close to the border with South Australia. It is also on the main east-west route between Adelaide and Sydney. (Photo: Graham Thompson)

SECTION 3
Power-Plus Range

There was a significant change underway in lorry buying patterns by the late fifties. This was influenced by several factors. Heavy duty rigids like Beavers were being usurped by lighter medium weight models such as Leyland's own Super Comet and their arch rival AEC's Mercury. Compared with a smiliar sized Beaver these were at least one ton lighter unladen, allowing that additional margin of payload. As such lorries were also cheaper to buy and run they were selling in large numbers. True, they might not have given quite the service life of a Beaver, but they were high quality vehicles in their own right. And in those years just as today maximising payload was essential. Unless an operator had to pull a drawbar trailer with his long wheelbase four wheeler it was more cost effective to chose a medium weight lorry.

Also the growth in articulated lorry usage was starting. Whilst Beavers had been available as dedicated tractive units from almost the inception of the model, such outfits were quite rare in many general haulage fleets. Leyland Motors' associate company Scammall of Watford ruled the roost with its Artic Eights which, for instance, were very popular with the major oil companies as tankers. At the light end of the market the Watford firm dominated urban and local distribution with firstly, Mechanical Horses, and then Scarabs. Both these little artics were sold in their thousands to railway companies. Articulation was also viewed with suspicion by many transport company owners who were wary of the reputation for jack-knifing of such lorries. When braking systems improved and were specifically designed for articulated tractive units such incidences became rarer. During the fifties orders for artics grew steadily, especially in the medium weight sector. However, in the 24 tons gross vehicle weight category the eight wheeler still remained popular. By the end of the decade there were definite signs that the days of supremacy of the eight wheeler were numbered. Top weight articulated lorries offered greater versatility by being able to use different trailers, thus giving better vehicle utilisation. Their maximum gross weight was identical to that of a four axle rigid, and maybe legal payload was a bit less, but other advantages were beginning to outweigh disadvantages.

Another consideration for designers and engineers was the dawn of the Motorway Age, coinciding with construction of improved and dual carriageway trunk roads. Quicker journey times were feasible if lorries could be speeded up. While the long-standing 20 mph heavy lorry speed limit had been abolished in 1957 not many heavyweights were capable of much more than 40 mph flat out. Things had to change with more powerful engines and additional gear ratios being necessary to take advantage of the improving road network. If lorries were to become faster they also needed to stop more effectively, so braking systems also had to be upgraded.

In September 1960 Leyland Motors announced a new Beaver range capable of meeting the demands of Motorways. All the main driveline components had been revamped and Leyland applied its Power-Plus label to its re-worked heavyweight lorries. Power-Plus Beavers were available with five wheelbase choices, including an 8 feet 0 inches tractive unit. All could be either left or right hand control and were rated at 14 tons gross weight solo, and 24 tons gross train weight for the tractor, or rigid with a drawbar trailer. Somewhat confusingly the model designations did not alter much from previously remaining at 14.B, but with a double figure suffix.

The established Leyland 0.600 and 0.680 engines were modified and re-tuned to become Power-Plus 0.600 and 0.680 respectively; sometimes referred to as P.600 and P.680 (although 'P' as a prefix had denoted 'petrol' in Leyland terminology when this fuel could be specified for a brief period in the late forties for their '600' engine). For this study the 'P' designation will be employed for Power-Plus units to prevent confusion with their immediate predecessors.

All the significant engine dimensions remained unchanged and the P.600 now produced 140 bhp at 1,700 rpm with 438 lb ft of torque at 1,200 rpm. It was often referred to as the economy engine option. The P.680 unit was uprated to greater effect producing 200 bhp at 2,200 rpm, which was one third more powerful than the 0.680 engine. Torque output was 548 lb ft at 1,200 rpm. Additional power had been achieved in several ways. Cylinder head design was improved and along with utilising twin inlet and exhaust manifolds this gave better gas flow. Polished connecting rods were used along with a different kind of cylinder liner in the P.680. Re-designed pistons which incorporated a new idea of combustion chamber in their crowns had been developed, known as the 'spheroidal' pattern. A CAV fuel injection pump (with hydraulic governor on the larger engine) was chosen with Leyland 4 hole multi-spray injectors.

The "bottom end" strength of Leyland engines had gained the company a reputation for remarkable longevity over the years. But an even better crankshaft was introduced into Power-Plus units. It retained identical journal sizes to earlier ones but was now forged from a steel and chrome-molybdenum alloy. This was hardened by nitriding and counterbalanced with a vibration damper to absorb torsional stresses. Home market and right hand control Beavers were powered by the P.600 engine as standard with the P.680 optional. This was the usual fitment for export and left hand control models.

Amazingly the long serving $16^1/_4$ inches diameter single plate clutch survived and its hydraulic operation was provided with air assistance for the P.680 engine. The 5 speed direct drive constant mesh gearbox could still be ordered but optional transmissions were now available. A straightforward 6 speed overdrive version of the basic unit; and this latter gearbox could also be specified for most chassis types with an additional deep crawler ratio. Nominally a 7 speed design the lowest gear was selected by a separate lever and was front mounted to drive the layshaft. A skilful and practised driver could use it to split the other 5 main gears, effectively doubling the ratios on hand.

The driveline was completed by a new rear axle. By then worm and wheel arrangements were rather dated so the traditional Leyland component was replaced by a heavy duty, double reduction, spiral bevel type. Primary reduction was by pinion and crown wheel with secondary epicyclic hub reduction. The standard ratio offered was 6.06:1 with higher and lower gearing options also available.

Marles cam and double roller steering remained but hydraulic power assistance was provided. The Beaver's suspension was softened and telescopic shock absorbers were fitted to the front axle. To cope with higher speeds larger brakes were utilised with wider linings front and rear. Those at the front were $4^1/_2$ inches wide with the rear shoes being 7 inches wide. Smaller diameter brake drums of $15^1/_2$ inches were used which were better for withstanding brake shoe pressures. In total there was 623 square inches of braking area. Actuation of the Westinghouse air pressure system was by diaphragm chambers and S cams. Automatic Bendix slack adjusters were also fitted. The parking brake operated on the rear wheels via a cross-shaft with air servo assistance for the pull-on handbrake lever.

An up-dated chassis frame was provided for all wheelbases. The three longest had $9^1/_4$ inches deep side rails with those of the specialised tipper being the same but with additional flitching at the main stress points. The tractive unit boasted a chassis almost 12 inches deep.

To complete a total revision of the Beaver's specification it featured an attractive modern cab. A couple of years previously Leyland Motors had begun fitting proprietary cabs on Comets and Super Comets. This structure was manufactured from pressed steel by Motor Panels of Coventry. It was also supplied to Dodge for their lorries as well as Leyland and its sister company Albion. Known as the L.A.D. cab (Leyland-Albion-Dodge) this design was now modified to permit deeper doors with entrance steps forward of the wheel arches. In turn this allowed the front axle of the lorry to be retracted. Leyland Motors' marketing department publicised this new driver's structure as the Vista-Vue cab, which with a large single piece windscreen gave improved vision from inside. It was mounted on flexible rubber supports for driver comfort although surprisingly heating and demisting equipment was optional and not standard. The Vista-Vue cab was not particularly heavy but a lighter version with glass fibre panels was also available. This variant was almost two hundred-weights less than the steel counterpart but it was more costly. Not many buyers of Power-Plus Beavers chose this option because of the higher price, and also doubts about its robustness. It was soon discontinued by Leyland Motors.

On 22nd September 1961 'Commercial Motor' published its results of a road test with a Leyland Power-Plus Beaver tractive unit. Their Chief Tester, John Moon was the adjudicator and the lorry was coupled to a Scammell tandem axle semi-trailer, 25 feet 7 inches long. This was just about the longest permissible then. Unladen weight of the complete outfit was rather disappointing at $9^1/_2$ tons, which allowed a maximum legal payload of $14^1/_2$ tons. This Beaver had a P.680 engine, 6 speed overdrive gearbox, and standard ratio back axle. Mr Moon reported that the engine pulled well with good torque characteristics at moderate revs. It performed well on the famous Parbold Hill which was always part of Leyland's Lancashire test route. A formidable climb, it is almost one mile long with an average gradient of 1 in 12, and a steepest section of 1 in $6^1/_2$. The Beaver took 4 minutes 20 seconds to complete the ascent and was in second gear, the lowest needed, for just over one minute. This was judged a good result and there was no noticeable exhaust smoke and very little rise in coolant temperature.

Descending the hill and the brakes were very effective, almost completely fade free according to the tester. On the M6 Motorway (Preston By-pass) the Leyland romped along at an average speed of 47.8 mph on a 17 miles run with 60 mph being recorded on certain sections. Overall fuel consumption was not recorded but analysis of the different results quoted for various tests suggest that this fully laden Beaver on typical mixed general haulage traffic and routes would return between 8 and 8.5 mpg; very respectable figures for the early sixties. The only criticism made by the driver was about high in-cab noise levels. In conclusion Mr Moon was very impressed with the Power-Plus Beaver and thought the ride good for a short wheelbase tractor. He praised the lorry's total performance as "outstanding, with very few other lorries capable of matching it".

Leyland Power-Plus Beavers were virtually new models in their entirety, and by 1960 14 years had elapsed since the last modernisation of the range. Tractive units did become very popular and sold well. Leyland did respond to complaints about noise levels endured by drivers and introduced sound deadening insulation. The other main grumble about the Vista-Vue cab was lack of space but vision was good with a comfortable driving position for drivers of average height and build.

L.A.D. cabbed Power-Plus Beavers were a completely new design, but maintained Leyland Motors' reputation for premium quality engineering. They were especially popular as tractive units as this concept rapidly gained popularity with heavy lorry operators. W.T. Flather Ltd. was a well-known manufacturer of specialist steel in Sheffield.

Preston Docks on a wet winter day could be surprisingly bleak for an inland river port. Northern Ireland Trailers Ltd were based there sending unaccompanied trailers across the Irish Sea by ferry nightly. This Beaver with its BTC 4-in-line trailer is typical of early 1960s road haulage.

The Roadrailer was an interesting concept that never caught on. The trailer was designed so that it could be used on standard gauge railway lines at a time when it was believed there could be greater inter-change- ability between road and rail transport. This Beaver tractor was used for demonstrating the principle.

Pochin construction used Leylands for many purposes, including moving their major items of plant by low- loader. Their Beaver is in charge of moving a Ruston Bucyrus 22-RB lattice jibbed excavator crane to Wales, with a Leyland Comet carrying the remainder of the jib. Pochin are still in business and have recent- ly built the new ERF factory at Middlewich.

As with 14.B models, Power-Plus Beaver tractive units for Austrialia had longer wheelbases than home-market lorries. McGrath Trailers used their own Beaver for transporting trailer chassis.

Deliveries of flour in bulk to large, plant bakeries commenced in the early 1950's when powder fluidising and pressurised discharge was perfected. This design of tanker was still common in the early sixties when the Power-Plus Beaver entered service. The tank did not tip, but had a sloping floor, and a "land-based', blower discharged it, so the lorry and trailer did not need to carry its own blowing equipment. S.C.W.S. were at Regent Mills, Glasgow (now demolished), just across the River Clyde from the U.C.B.S. Bakery. The building on the right is Kelvin Hall, former home of the Scottish Commercial Vehicle Show, and it is now the site of the City's Vehicle Museum.

Leyland Motors had reasonable levels of sales into Europe after the Second World War until the seventies. This Swiss Beaver has had a sleeper extension fitted to its cab for trans-European journeys, and it has been bodied with the ubiquitous continental tilt bodywork.

Norway was the home country of this Power-Plus Beaver liquid gas tanker and trailer. It is a right hand control lorry, and has interesting sized front tyres fitted, very similar to "super-single" sizes fitted today by some operators on tractive units. This is once again proof that there is very little totally new in road transport.

A late Power-Plus Beaver, registered in 1965 and compliant for the new C. & U. Regulations, with a 33ft long trailer, Fleetwood Fish Transport was founded in 1965 as a co-operative venture between the Port Authority, trawler owners, and fish wholesalers when British Railways closed the town's railway line. Several depots were established and nightly trunks were run to Billingsgate, London, and other large city's fish markets. The 8ft containers were ex-railways, and were insulated with dry ice for maintaining a cool temperature.

Hunter's of Hull are another well remembered transport company, and they were early European hauliers, establishing TIR services in the early sixties. Their 1965 Beaver is coupled to a 33ft long Crane Fruehauf tri-axle trailer. Don't be fooled by the relatively few reinforced steel drums; there is a full twenty tons on board. The Ro-Ro ferry Norwave is in the background.

Certain kinds of indivisible loads have always been exempted from normal C. & U. Regulations, without the need for police escorts or heavy haulage permits. Steel girder haulage falls into this category, and in the era of Power-Plus Beavers, extendable (commonly known as trombone), trailers of up to 45ft could be used. McFaddens of Rutherglen, Glasgow, used such an outfit with their Beaver, photographed on 21st December 1966.

J. & A. Smith of Madiston had grown into one of Britain's biggest haulage companies by the mid-sixties. There were depots throughout Scotland and England. This Power-Plus Beaver is beginning to show signs of several years hard work and it is coupled to a canvas topped, Dutch TIR trailer. A large proportion of Smith's fleet, which numbered about 500 lorries, were Leylands.

Petroleum companies were good customers for Leyland Motors for many years, and Leylands were popular for airport refuelling duties. This Beaver was in service at Don Muang, Bangkok, as part of the aircraft refuelling fleet of tankers belonging to Esso. It has a strange tank; note the corrugated front bulkhead, and large discharge valves.

A close coupled Power-Plus Beaver and livestock trailer in service with Atkinson's Ltd. of Tallygarapoona, which is a quite small country town north of Shepparton, Victoria, Australia. A very smart outfit and obviously well cared for. (Photo: Graham Thompson).

Aboods Transport of Sydney ran a mixed fleet of marques through the years including Leylands. Power-Plus Beavers featured on their inter-state haulage runs and they were fitted with sleeper cab extensions and an extra fuel tank. In keeping with company practice this lorry was named "Paula Maree", a double Christian name taken from two members of the extended Abood family. (Photo: Len Bartlet).

It's a tight squeeze under this bridge for the Beaver operated by Australian transport giant, Brambles. It was almost at the end of its journey from Eglo Engineering at Silverwater, Sydney, to ICI Australia's chemical plant at Botany. It can be safely assumed that the lorry was about to come to a halt as the driver operated mechanical hand signal is in the partially extended position. (Photo: Young & Richardson, from Tony Petch)

This driver is about to take delivery of a new Power-Plus Beaver and refrigerated trailer on behalf of Streets Ice Cream at the company's Turrella factory in Sydney. In the sixties it would appear that it was common pratice in Australia to suspend the refrigeration engine and compressor under the trailer, rather than mount it on the front bulkhead. (Photo: Young & Richardson, from Tony Petch).

SECTION 4

Freightline and 'Two-Pedal' Beavers

Power-Plus Beavers remained in production for a relatively short time. The 1960s was a hectic period with, amongst other considerations, far reaching legislative changes dictating lorry designs and operating conditions. Not only did Leyland Motors need to react to amended Construction and Use Regulations which increased gross vehicle weights and lorry dimensions, the company itself also expanded out of all recognition as the decade advanced. This growth was by acquisition and followed a pattern set by the takeovers of Albion Motors and Scammell during the fifties. Both these famous and respected firms had settled quite comfortably into the Leyland Motors empire. But the merger in 1962 with the ACV Group was a different matter. Leyland was the dominant partner in an amalgamation of two large businesses with contrasting management styles. Consequently, and particularly as time passed, they found it very difficult to work harmoniously together.

In the year of 1964 Leyland Motors announced a Freightline range of lorries which encompassed most of their model catalogue. The striking main feature was revealed with a totally fresh approach to cab design. Freightline Beavers were based mechanically on their immediate Power-Plus predecessors but now had higher gross weights to take advantage of the 1965 Construction and Use Regulations. Solo Beavers were rated at 16 tons gross vehicle weight and tractive units at 30 tons gross train weight. Because articulated outfits now had a substantial payload advantage over rigid eight wheelers the demise of such lorries for general haulage duties was hastened. One additional Beaver chassis, a 10 ft 0 inches wheelbase model, was introduced into the range. From then onwards the vast majority of Beavers purchased were articulated tractive units.

Freightline Beavers were offered with identical driveline specifications and options as Power-Plus models. The only significant difference being the availability of close ratio versions of the 5, 6 and 7 speed gearboxes. Also, an alternator replaced the dynamo. Initially this was a CAV component. The braking system remained unchanged apart from the provision of a three line circuit for the tractor. This was a legal requirement of new regulations.

New chassis type designations were introduced: 16BT etc (14BT for tractive units) with the number indicating gross vehicle weight and 'T' denoting tilting cab. Most Beaver variants were available with right or left hand control driving positions. The completely new cab was a revelation and a tremendous advancement for the driver. While it did revert to a very angular visual image, whereas L.A.D. cabs were more rounded, it was much quieter and more comfortable inside than anything made before.

Leyland named this new driver's structure the Ergomatic Tilt-Cab, and it tilted forward 55 degrees to allow access to the engine. It was a welded double skin steel fabrication on box section pressings made by Joseph Sankey (later GKN Sankey) at Wellington. With the lower entrance step mounted ahead of the wheel access was easy and a large single piece windscreen afforded excellent vision. The driving position was first-class with all switches and controls positioned within very easy reach of the driver. Instruments were grouped together in a neat binnacle with one quick glance sufficing for the immediate conveyance of information.

Comfortable, fully adjustable seats were fitted and with copious insulation internal noise was quite low. Powerful heating and demising equipment was provided as standard. The studios of Giovanni Michelotti, a well known designer of the period, were responsible for the overall design of the cab and it became an instant success. However, it did have some shortcomings. Interior space was still not plentiful as a large engine cover intruded. Because this was slightly offset from centre, left hand control versions were extremely cramped for the driver. As time progressed it became apparent that the angular construction with hollow box sections were not conducive to longevity and premature corrosion became common.

The main development of Freightline Beavers was the introduction of semi-automatic gearbox versions and these are studied in detail later in this section. Minor amendments to manual transmission types were made in 1968 with P.680 powered tractive units being uprated to 32 tons gross train weight. Further legislation had permitted longer trailers which gave greater axle loading tolerances. It was also decreed that lorries must have a minimum 6 bhp of power per ton. Another refinement was the incorporation of a load sensing valve into the braking system. The other change of importance was the replacement of CAV alternators, starter motors, and ancillary equipment by BUTEC components. BUTEC was a Leyland Group company established to manufacture such items using designs licensed to them by Leece-Neville of Ohio.

In 1970 a severe rationalisation of the Beaver range occurred as Leyland's totally new '500' series Buffalo was launched into the market. Manual gearbox Beavers remaining were a long wheelbase load carrier and a 9 ft 6 inches wheelbase tractive unit. The rigid lorry had a 5 speed wide ratio constant mesh gearbox as standard with the option of a 6 speed close ratio unit. This was standard for the tractive unit. This latter model could also be specified with a more powerful turbo-charged engine developed from the P.680. Known as the Leyland 690 engine it will be described in detail shortly.

Two-Pedal Beavers

Even in the mid-sixties the standard gearbox for British built premium lorries was typically a constant mesh type. They had 5 or 6 gears and as revealed earlier Leyland did have a 7 speed unit which could be used in splitter mode to provide additional ratios. By then some medium weight and many lightweight models had synchromesh transmissions but the conventional constant mesh gearbox still reigned supreme for heavyweights. Such units were usually robust and trouble free, but effective operation was dependent upon the driver's skill and the use of a heavy clutch pedal.

Leyland Motors caused a mild sensation when they announced a semi-automatic version of their Beaver tractive unit at the 1966 Commercial Motor Show. Their introduction of such a version, or 'Two-Pedal' Beaver as it became known, was bold and innovative.

As a leading passenger vehicle maker Leyland Motors had many years experience of pre-selector and semi-automatic transmissions. Pre-selector gearboxes still retained a third pedal (which was the gear change mechanism) and had been used in buses since the late twenties. Semi-automatic designs came a few years later. In 1960 Leyland Motors acquired Self-Changing Gears Ltd of Coventry, one of the foremost producers of such systems. So it was a logical step that a semi-automatic lorry would become available in due course.

The first 'Two-Pedal' Beavers entered service by the end of 1966. This model was a standard specification unlike manual gearbox versions which offered engine and transmission options. Designated 14BT.28R, semi-automatic Beavers were powered by Leyland's P.680 engine exclusively.

The wheelbase was 10 ft. The Beaver's 5 speed gearbox was based on the passenger chassis semi-automatic unit made by Self-Changing Gears Ltd. This consisted of a compound series of epicyclic gear trains and a multi-plate clutch for direct drive. The gearbox was unit mounted with the engine with drive from the flywheel being taken through a fluid coupling combined with a centrigual lock-up clutch. Soon this transmission would become known as the Leyland Pneumo-Cyclic. Whilst a 5 speed direct drive unit was perfectly acceptable for passenger vehicles, more ratios were desirable for lorries operating at 30 tons gross train weight. To achieve this a special Leyland rear axle which incorporated an Eaton 19,800 two-speed driving head was produced. This was a fully floating spiral bevel single reduction unit. Its usual ratios were 4.87:1 (high) and 6.63:1 (low). Alternatively, 5.43:1 and 7.39:1 gearing was available. With the standard back axle a comfortable top speed of 60 mph was achievable.

To make a gear change the driver operated a pedestal mounted lever through a gate change. This automatically released the brake band of one gear train and applied another. Actuation of brake bands was by air pressure and of course there was no clutch pedal. Back axle ratio changes were accomplished by simply flicking a switch mounted on the main gear level and momentarily releasing the accelerator. With practice drivers soon became proficient at making full ratio, half ratio, and split changes rapidly. An inexperienced driver could "lose" the back axle ratios if he tried to be too ambitious too soon, but the techniques required were easily mastered by keeping it simple to begin with.

Hard nosed and sceptical transport bosses are usually wary of any vehicle as revolutionary as a semi-automatic Beaver was. Leyland Motors were confident they had a winner so they dedicated a new production line solely to the assembly of these lorries. This faith was rewarded and by March 1967 in a little over 3 months, over 250 had been sold. *'Commercial Motor'* published a road test of a 'Two-Pedal' Beaver in April 1967. Their stalwart tester A.J.P. Wilding conducted the appraisal and was very impressed. He believed the semi-automatic gearbox was worth an additional 15 bhp when compared with a manual Beaver. Also, acceleration times when fully laden at just over 30 tons gross weight were appreciably quicker. Perhaps most surprisingly fuel consumption was slightly better, together with a higher average speed obtained over the normal model. The only slight disadvantage was the loss of about 3 cwt of legal payload resulting from the heavier semi-automatic gearbox installation.

A development of the semi-automatic Beaver came along later in 1967 with the incorporation of an integral splitter unit into the gearbox. This was a "step-up" design to provide an approximate 50% split between each of the five main ratios, which were effectively doubled to provide 10 speeds. The highest gear was a true overdrive with a ratio of 0.77:1. This splitter unit comprised an air operated epicyclic gear train mounted on the front of the main gearbox casing. Again this 'Two-Pedal' model was standardised and provided with a new type designation – BE.TR. 1PR. The wheelbase was reduced slightly to 9 feet 6 inches and the lorry was rated at 30/32 tons gross train weight. The higher weight only being possible then in 1967 with a tri-axle trailer. (This was to change a few months later in 1968 when overall length restrictions were relaxed). Final drive was now from the Leyland heavy duty spiral bevel hub reduction back axle. A choice of ratios was available, the standard being 6.06:1. There was a lower option of 7.74:1 and higher choice of 4.82:1. Any purchaser specifying the latter would really have had a flying Beaver, with a theoretical top speed of about 75 mph.

Until the splitter gearbox 'Two-Pedal' Beaver won full acceptance the earlier two-speed axle version remained in production. In 1968 both models became rated at 32 tons gross train weight. Later in the year these tractors were rationalised when the 10 speed Pneumo-Cyclic gearbox became standard. Several other modifications were made to the tractive unit and an exciting engine option was announced. This was a turbo-charged version of the 0.680 unit, becoming known as the Leyland 690 engine.

Leyland Motors had years of experience with turbo-charging having investigated the principles for rail-car engines in the late thirties. Their innovative '500 Series' power units, also announced in 1968 but not becoming available until several months later, also had turbo-chargers on some derivatives. The Leyland 690 engine retained many common features with the P.680 unit but there were important differences apart from a Holset turbo-charger driven by exhaust gases. The cylinder heads had larger inlet and exhaust valves; both the exhaust and air inlet manifolds were re-designed; and the pistons were oil cooled. A larger capacity fuel injection pump was fitted. This had a feature which controlled the fuel pump rack according to the inlet manifold pressure. This device prevented excessive fuel (and black smoke) being injected when the engine was quickly accelerated from low speed.

'Two-Pedal' Beavers could now be ordered with either the normally aspirated P.680 engine or the turbo-charged Leyland 690. Yet another series of type designations was used:– BV69/32 PTR for the latter, and BV68/32 PTR for the former. The Leyland 690 engine produced 240 bhp at 2,200 rpm with 650 lb ft of torque at 1,400 rpm. These values were to the BS AU141:1967 scale, but at that time "metric" or SAE ratings were beginning to be used. Under that formula the corresponding output was 258 bhp and 700 lb ft of torque. The wheelbase of 9′6″, final drive type and ratio options were identical to the preceding BE.TR.1PR model. Other specification changes for this new Beaver included a revised parking brake system utilising a spring brake concept on all four wheels. Spring brakes work on a "stored energy" principle to give a fail-safe facility. If air pressure falls below a certain level, (usually 60 lbs per square inch), powerful coil springs in the actuators overcome air resistance and apply the brakes.

At the time of announcement of the Leyland 690 powered Beaver virtually every other British made top weight tractive unit on sale had around 180 to 200 bhp available at the most. An additional 40 bhp was a very attractive proposition and the only other home produced tractor with similar power was AEC's Mandator V8 at 247 bhp. (See Commercial Vehicles Archive Series book 'AEC Mandator V8'). What an exciting prospect the turbo-charged 'Two-Pedal' Beaver was. By any yardsticks applicable then it was a powerful lorry with a simple, clutch-free gear changing method which any driver would relish.

By 1970 British Leyland (as Leyland Motors had become) was strongly marketing its '500 Series' lorries which included the Buffalo, a new top weight tractive unit with manual range-change 10 speed gearbox. As discussed earlier in this section the entire Beaver line-up was rationalised to just a long wheelbase haulage model and normally aspirated and turbo-charged 'Two-Pedal' tractive units. Even so, the end for semi-automatic transmission was in sight as both models could be specified with optional close ratio 6 speed constant mesh gearboxes.

It is believed that almost 600 'Two-Pedal' Beavers of all variants entered service. At that time the concept was not adapted by any other main-stream lorry maker and Leyland would have been years ahead of their competitors if it had been successful. The first version with basic 5 speed gearbox and 2 speed rear axle was least troublesome although fluid flywheel gland failures were common. This was a persistant fault with all forms of the transmission. Ten speed splitter gearboxes were far more complex which resulted in additional problems. Turbo-charged Leyland 690 engined Beavers suffered from cylinder head gasket failures. Over-heating was also experienced and the Ergomatic cab design contributed to this by constricting air flow around the engine and restricting radiator coolant capacities. The higher power and torque outputs of the Leyland 690 (and also AEC V8 engine when mated to this gearbox) could cause the gearbox oil to actually boil after long, high speed motorway running, giving rise to serious internal damage.

Despite the additional costs of a semi-automatic lorry and suspicion by workshop managers and fitters of a complex clutch and gearbox, Leyland Motors achieved commendable sales figures for 'Two-Pedal' Beavers in a short period of time. Whilst the concept was popular and reliable for passenger vehicles, the extra demands made by heavy lorries found the transmission wanting. The idea had tremendous potential but ultimately failed because of lack of development. This was a common theme shared with other Leyland group projects at the time. British Leyland, the parent company, was entering its nadir by becoming involved with volume car production. This brought political interference, industrial relations unrest in the car divisions, and crippling financial losses by car producing factories. The profitable heavy vehicle divisions had to subsidise the whole group and quickly became starved of development and investment funds. The timing could not have been worse as Scandinavian and European lorry makers made rapid in-roads into the British market.

So, 'Two-Pedal' Leyland Beavers became lost causes of the British Leyland empire. This was a great pity because Leyland was so far ahead of its competitors in pioneering such a concept. Even today, over 30 years later manual gearboxes are still fitted by the majority of heavy lorry builders. Yet in the sixties Leyland was pioneering an easy to use, fool proof transmission system which still allowed the driver full control of gear selection options. Sadly it was not to be, but let us at least recognise this attempt to make a lorry driver's job easier.

By the end of 1971 production of all Beaver tractive units was being phased out as Leyland pinned its faith on the disastrous (as it proved later) '500 Series' Buffalo. A few batches of long wheelbase Beavers were occasionally produced during the following couple of years. These were usually destined for the Ministry of Defence and some of the major oil companies who favoured the model for drawbar fuel bowser duties at military airfields and civilian airports. By the end of 1974 the illustrious Beaver had been discontinued after a production run spanning 46 years, making it one of the longest lived lorry types of the twentieth century.

Although the Leyland Freightline range of tilt-cabbed models was announced in 1964, it was late in the following year before any became available. This is a very early Freightline Beaver and it entered service with J.W. Lees, an independent brewer from Middleton Junction, Manchester.

The transport legislation of the mid-sixties transformed the industry and opened the floodgates for articulation. Pointers Transport used a small fleet of Freightline Beavers with specially constructed step-frame trailers for carrying pre-fabricated concrete building sections for Taylor Woodrow.

D.T. & N. Keighley Ltd., Dutton Transport, of Burnley placed this early Freightline Beaver in service in 1965. It was photographed loading glassware at Mullards Glassworks, Padiham. This lorry has survived to the present day and is beautifully restored in the livery of Reids Transport of Minishant, Ayr.

Bridge's Freightline Beaver is about to depart from its St. Helens base with a load of glass bottles for Busby, between East Kilbride and Paisley. This particular Leyland was used on night trunking duties in the sixties, the usual changeover points being at the famous Jungle Café near Shap, or the Kirtlebridge Café.

A fine study of old and new Beavers, with the striking difference between the cab styles evident. Freightlines were mechanically very similar to Power-Plus versions, with the main differences being in braking systems that were compliant with the 1965 C & U Regs for the tilt-cabbed lorries. Brakes on the L.A.D. models could be easily upgraded to meet the new minimum standards. These tankers were loading at the Mobil Refinery at Coryton.

International Alloys Ltd. purchased this Beaver articulated outfit in 1966 for transporting flasks of liquid sodium. The single axle step-frame trailer is unusual for such an application, and note how instructions for dealing with fire are displayed on the trailer's side raves.

By the mid-sixties, as a result of operator demand, most four wheelers were medium weight models, which had lighter unladen weights than heavy-duty lorries, but offered the same maximum gross weights. If a trailer was being pulled, the more powerful heavyweight was then preferred. For normal road going work Beaver tippers by 1966 were becoming rare, and this aggregate bodied example entered service with Webbs of Cambridge, who are still in business today.

Ready to leave with a load of R.White's lemonade, and the Beaver is coupled to a specially designed trailer with higher headboard than usual. This allowed double stacked pallets of soft drinks to be supported, and the sheets were also made to the exact dimensions of the load.

The introduction of Freightline Beavers in 1965 saw Leyland offering a choice of wheelbases for tractive units. Restrictive overall length allowances pre-1965 determined the common tractor wheelbase of 8 feet for home market models. Post 1965 legislation allowed longer articulated combinations, and a 10 feet wheelbase tractor became available. Lyons stayed with the traditional 8 feet wheelbase for their early tilt-cabbed Beaver.

British Industrial Sand Ltd. preferred a longer wheelbase of 10 feet for their Beaver artics. This permitted the blowing equipment for discharging the tank contents to be more easily mounted on the chassis. The tank is a bottom discharge, non-tipping design and in concept and shape is identical to those now very popular with bulk powder carriers. Remember, this photograph was taken over 30 years ago, and we can safely state once more, there is nothing new in road transport!

Another unusual trailer is behind Nugent's Beaver. It is long for a tipper of its era, and is low sided for aggregates. Note the single tipping ram, with a massive hydraulic oil tank behind the cab. J. Nugent operated from Brock, near Garstang, and this delivery was being made at Quick Mix, Upholland, between Wigan and Skelmersdale. The crushed stone was from a Carnforth quarry, and Grisdales of Churchtown, Garstang, made the trailer bodies.

The 1965 C & U Regs. spawned plenty of innovation in the sixties. Longer overall lenghs and higher gross weights allowed greater latitude for operators and designers. A tri-axle trailer such as this gave the maximum 32 tons gross vehicle weight and B.I.C.C. worked with Carmichaels in designing a vehicle specific for their needs. The twin tanks allowed different coloured PVC pellets, used for the outer covering of electrical cables, to be delivered. The Beaver and trailer were being demonstrated at B.I.C.C. Helsby, prior to the outfit entering service.

Articulated mobile concrete mixers have never been popular in this country; the rigid six wheeler reigning supreme for such applications. One or two concrete suppliers have tried artics, such as this Beaver and mixing drum mounted on a 4-in-line trailer. Hobbs was in the Bristol area.

A double trailer combination headed by a Beaver, in service with Caltex in New Zealand, comprising a standard fifth wheel semi-trailer and second semi-trailer on a drawbar. The Beaver had a six-speed gearbox plus deep crawler, to give 10 ratios, and it was photographed at Whangarei, North Island, in January 1980. (Photo: Rufus Carr).

This fine looking Beaver entered service with Total Oil Australia in 1968. It was mainly used for fuel deliveries in the Sydney metropolitan district and was photographed at Total's former depot at Moore Street, East Botany. (Photo: Young & Richardson, from Tony Petch).

Sydney County Council (now Energy Australia) bought this Beaver in 1969 for low-loader duties. It has a substantial crew cab and chassis mounted winch. It worked mainly in the Sydney area. (Photo: Energy Australia Archives, from Tony Petch).

Leyland Motors had a long history of supplying fire-fighting appliances, and all Beaver models down the years were used in this demanding role. Tilt-cabbed Beavers were no exception and City of Plymouth Fire Brigade placed in service this hydraulic platform equipped machine. By the late sixties such equipment was replacing traditional extendable ladders.

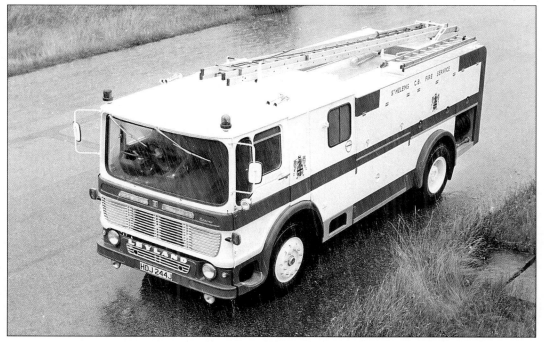

The north-western based Fire Brigades were keen Leyland buyers, and St. Helens Fire Service introduced this Beaver based appliance in 1971. It was fitted with a Pneumo-Cyclic semi-automatic gearbox.

Leyland Motors really should have had a winner with the 'Two-Pedal' Beaver concept. It offered the driver clutch free gear changing with full control of all the ratio options. Robinson's Transport of Hereford took delivery of this 14BT.28R version (two speed axle), in 1967 and it represents a typical general haulage lorry of its time.

Fleetwood Fish Transport also had 'Two-Pedal' Beavers in its predominantly Leyland Fleet. Usually, these Beavers had wider front wings than manual gearbox versions, and this was a useful aid to spotting one without having to peer into the cab to examine the gear stick.

In 1968 a revision of the C & U Regs. permitted 40 feet long trailers in Great Britain. This was intended mainly for ISO container hauliers, but the entire transport industry welcomed this move. Here steel stockholders Miles Druce have a new forty footer behind their 'Two-Pedal' Beaver, and this is a BE.TR.1PR version, with integral splitter unit mounted on the gearbox.

The decision to turbo-charge the P.680 unit to the Leyland 690 engine gave the company the chance to compete with the influx of Scandinavian tractive units in the late sixties. This side view of a BV69.32PTR Beaver shows the large air cleaner, mounted behind the cab. With its semi-automatic gearbox this was the ultimate 'Two-Pedal' Beaver, but sadly, lack of consistent reliability caused the model's downfall.

SECTION 5

Beaver Derivatives

Leyland Steer – Twin Steering Six Wheeler

As the British commercial vehicle industry evolved it created certain kinds of lorry rarely produced elsewhere in the world. This resulted from somewhat inhibitive Construction and Use Regulations coupled with restrictive carriers licensing legislation. Even the famous rigid eight wheeler was born from a quirk of the legislation enacted in 1933. This particular configuration remained almost unique to Britain and its former Empire and Commonwealth states which imported our lorries. The well remembered 'A' Licence, a permit greatly coveted by general hauliers was granted with a weight allowance. This determined the size of lorry an operator could use. For many years the amount of road fund tax payable annually was calculated according to vehicle weight and engine horse power.

By the mid-thirties haulage contractors were exploring every legal (and sometimes illegal) means possible of achieving maximum payload from available lorries allowed by their 'A' Licences. There was quite a demand for four wheeler to six wheeler conversions by the addition of a trailing third axle. Several engineering firms specialised in such adaptations which made a lighter lorry than a factory built conventional six wheeler. However some of the conversions were rather crude. They had braking and suspension components which were poorly matched with those of the original vehicle.

Shortly after the first eight wheelers entered service in 1934 some heavy lorry builders introduced twin steering six wheeler models. It could be assumed that these were based on four axle chassis but minus the rearmost axle. But it would be incorrect to do so. To begin with Construction and Use Regulations decreed that for a six wheeler all three axles must be braked but there was no requirement for the second steering axle of an eight wheeler to have brakes. Because the gross vehicle weight of a twin steering six wheeler was less than that of a conventional three axle lorry, but greater than that of the largest two axle model, such designs were therefore based on heavyweight four wheeler chassis.

From the operator of such a lorry's viewpoint, if there was some margin to spare on the 'A' Licence, but insufficient for a standard six wheeler, a twin steering six was ideal. For less than half a ton of extra unladen weight a gain of about three tons of legal payload could be made. Also, the cost was also lower than for a conventional three axle model.

Towards the end of 1936 Leyland Motors announced their first purpose built twin steering six wheeler. As the company had already adapted its animal family model names it was logical for them to call their new lorry Leyland Steer. While this was the (American) name for a young male calf, it also conveyed perfectly the layout of the lorry. Available from 1937 the Steer was based on the forward control Beaver 6, but with a gross vehicle weight of $15^1/_2$ tons it could legally carry a payload of about $10^1/_4$ tons. Steers were made with a choice of two wheelbases, usually for platform or tanker duties. They could be powered by either 6 cylinder petrol or diesel engines. These lorries were suitable for drawbar trailer work with a gross train weight of 22 tons. Incidentally this was the same gross weight as an eight wheeler which at that date could not legally tow a trailer.

Leyland Steers were also offered with a couple of interesting options, similar to those of Beaver 6 tractive units. A 5 speed constant mesh gearbox could be specified which gave a choice of either direct drive and extra deep crawler ratio, or an overdrive design of 0.83:1 ratio. Also, like the Beaver 6 tractor, there were optional worm and wheel or spiral bevel double reduction rear axles. Each unit offered a selection of final drive gearing. For the Steer's brakes the Beaver's basic system was utilised, except those on the second axle were operated by a cable linkage from the master servo to a cross-shaft. The parking brake worked on this axle in addition to the back one. Steering of the second axle was by a fore and aft drag link.

If correctly loaded to its maximum a Steer would be almost perfectly balanced with a 50/50 split of weight between the front axles and rear. This gave better weight distribution than an ordinary six wheeler and of course, there was no tyre scrubbing which was a common problem with many three axle lorries. Such a twin steering six wheeler became known as a "Chinese Six" amongst lorry men. It is not known precisely how this terminology originated but one school of thought suggests that because anything non-conformist is often likened to a "Chinese way of doing things" this is how the name came about.

Leyland Motors suspended manufacture of Steers during the second world war but the model was re-introduced in 1947. Its mechanical specification was similar in all respects to that of the 12.B Beaver, with the same options. Any significant changes made to Beavers through the years were applied to Steers. With the increase in gross vehicle weights in 1955 the Steer was uprated slightly to $16^1/_2$ tons for a solo lorry and 24 tons gross train weight if pulling a trailer. By the end of the fifties demand for heavyweight Chinese Sixes had gradually dwindled and the model was discontinued, not featuring at all in the Leyland Power-Plus range. Just as the market for solo Beaver four wheelers had been annexed by medium weight lorries, so the same fate befell Steers. A medium weight tractive unit and semi-trailer offered a better payload with more flexibility. Throughout its entire period of production Steer sales had never been huge, but the model sold in steady numbers annually, fulfilling a special requirement for certain operators.

Nevertheless, even though the model had been discontinued in 1959 the type and name re-appeared a few years later in 1966. Just as the Construction and Use Regulations of the thirties had provoked the introduction of new designs, so history repeated itself with the 1965 legislation. These revised statutes strongly favoured articulated lorries with a maximum 32 tons gross train weight being legal on either four or five axles. However, in practice it was virtually impossible to obtain the maximum weight on four axles. This was due to the legal constraints imposed in respect of overall length, distances between the first and fourth axles, and individual axle loadings. The longest semi-trailer legally permitted was 33 feet and to qualify for 32 tons gross train weight with a four wheeler tractive unit the rear bogie had to be at the extremity of the trailer. This meant it was very difficult to get enough weight onto the bogie and conversely it was easy to overload the tractor's axles. The only operators who could achieve correct loading were those typically engaged in steel bar or girder haulage who were able to position load bearing supports where required. General hauliers needing maximum gross weight had two options: either a tri-axle trailer or a six wheeler tractive unit.

While the tri-axle trailer was the cheaper alternative there was a demand for six wheeler tractive units, and a twin steering type was preferable to a tandem rear axle design. In particular bulk liquids, powders, and liquefied gas hauliers favoured them as they believed the stability of the whole outfit was improved. To meet this need, in 1966 Leyland announced their Freightline Steer tractive unit. It featured lightweight front axles and was a very compact machine. The wheelbase of just 10 feet meant that the fuel tank had to be positioned transversely across the chassis because there was no space for it on the frame. Adwest type steering gear was fitted which could be power assisted if desired. The Freightline Steer was powered by a Leyland P.680 engine and had 'Two-Pedal' transmission. This comprised a 5 speed semi-automatic gearbox and two-speed rear axle. Brake shoes on both front axles were 3 inches wide and spring brakes were utilised for the parking brakes on the second and rear axles. In the interests of achieving a low unladen weight 9.00 x 20 inch tyres were fitted on all four front wheels with 10.00 x 20 inch covers on the back wheels. Leyland Freightline Steers were only in production for a couple of years because in 1968 the Construction and Use regulations were revised to allow 40 feet trailers. This meant that sensible outer axle distances could be achieved which met the 32 tons gross train weight criteria.

Leyland Super Beaver

Leyland Motors was famous throughout the world for commercial vehicles and was a prodigious exporter. Not only were all its mainstream home market models sold overseas, certain specials were produced to cater for specific needs abroad. Leyland's Super Beaver was an export version of the Beaver. This model was first introduced in the late forties in normal control, bonneted layout. Built around the standard lorry's driveline Super Beavers had bigger and stronger chassis frames, typically 12 inches deep. This made them extremely robust and durable and three wheelbases were available; a long wheelbase rigid for drawbar trailer work, a shorter wheelbase heavy duty tipper, and a tractive unit. All could be ordered with either left or right hand driving control. Gross weights ranged from 12 to 16 tons for solo lorries and was determined by the size of tyres fitted. Super Beavers could cope with heavier loads than found at home and in many countries regularly hauled massive burdens.

To better handle arduous operating conditions abroad Super Beavers were fitted with larger capacity radiators. A 2 speed auxiliary gearbox could be included into the transmission, arranged with either step-up or step-down gearing. Usually CAV fuel injection pumps were fitted to engines of exported lorries. The cab style remained virtually unchanged for the production life of Super Beavers. Incorporated into the model were all the improvements made to ordinary Beavers down the years. Super Beavers made from 1960 had Power Plus P.680 engines fitted. The model remained in production until 1980 and in latter years its uncomplicated specification made it an ideal lorry for third world countries.

Leyland Beaver Auto-Tractor and Auto-Articulator

In the year of 1950 Leyland Motors licensed The Auto-Mower Engineering Co Ltd of Norton St Philip, near Bath, to modify 12.B Beavers for the forestry industry. The first specialist model produced was a timber tractor based on a Beaver chassis of 10 feet $3^1/_2$ inches wheelbase. Designed for timber extraction work off road an Auto-Winch and anchors were fitted, along with a jib crane of 30 cwts capacity. The winch was driven from a power take off mounted on the gearbox which was specially adapted for the unit. To complete the outfit for timber haulage an Auto-Trailer pole carrier with bolsters was available.

The Beaver Auto-Articulator was designed specifically for round timber haulage. It was a 12 feet 6 inches wheelbase 12.B Beaver fitted with a combined turntable and bolster mounted directly above the back axle. The rear carriage of the semi-trailer was connected to the tractor by a swan-neck heavy gauge steel tube. The distance of the pole trailer's axle carriage from the tractor was variable to accommodate different sizes of logs. The maximum overall vehicle length was 34 feet, but it was quite common for round timber hauliers to have substantial over hanging loads.

Other Engine Applications

As a major commercial vehicle maker Leyland Motors had substantial engine manufacturing capacity. In addition to meeting its own needs the company was able to supply power units to competing lorry builders as well as various plant and machinery firms. Associate Group companies also used Leyland engines after the second world war.

Of course engine tyres fitted into early Beavers were also used for other Leyland models. Most notably these engines were fitted into Octopuses, Hippos and contemporary passenger vehicles. This practice continued in the post-war era and as well as the lorries mentioned 0.600 units were used in Leyland Titan PD1 and Leyland Tiger PS1 chassis. When the 0.680 design came along in 1951 it immediately became the favoured unit for multi-axle lorries Octopus, Hippo and Steer in addition to Super Beaver. Usually de-rated it was also fitted in PD2 and PS2 buses and coaches. Both 0.600 and 0.680 tyres were made in horizontal format for Royal Tiger coach chassis and railcar applications. Other lorry builders to fit these particular Leyland vertical engines were Scammell, Albion, Bristol, Guy and Foden. Different versions were also provided for many, varied industrial and marine purposes.

Such was the impressive reputation won by Leyland products and particularly the 0.600 unit that some foreign companies sought engine technology from Leyland Motors. DAF of Holland built 0.600 and 0.680 engines under licence and eventually its own designs were derived from the Lancashire product. Similarly Scania Vabis of Sweden used 0.680 know-how bought from Leyland. Other organisations such as WSK of Poland, Ashok of India and Pegaso of Spain utilised basic Leyland designs for their own developments.

There is no doubt that Leyland's 0.600 engine was world class. In addition to being a superlative lorry unit it was extremely versatile for powering other machinery. The more powerful 0.680 unit was based on its smaller predecessor and was also a fine engine. Maybe it was not quite as reliable, and especially so in P.680 guise, but it still did very well. Not only were these Leyland engines very effective and gave long operational lives, they were also quite frugal with fuel giving commendable economy.

Twin-steering six wheeler rigid lorries cornered a niche market from their inception in the mid-thirties for a period of approximately 30 years. Leyland Steers were often used by cattle hauliers for increased stability. When Sinclair's Steer and trailer was photographed in 1937, eight wheelers were not permitted to tow a trailer, so this was just about the biggest type of combination allowed in Great Britain for normal haulage duties.

Ansells Brewery of Birmingham was a committed Leyland user and there is a fair load of barrels on their post-war Steer and trailer. It's a nice sunny day and the second mate has chosen to ride on top of the load.

Leyland Steers were often found with drawbar trailers attached and this 15.S/1 is loaded to its legal maximum of 22 tons gross weight with sacks of cattle feed. Press, Bly & Davey Ltd. were based at Bidford-on-Avon in rural Warwickshire. The short chassis overhang behind the rear axle is apparent on this side view, which made it difficult to overload the back axle, and if loaded correctly, optimum axle loadings were obtained on the lorry.

There is plenty of atmosphere of the fifties in this photo taken at Preston Docks, loading pulp. It was standard practice at the time to put your sheets on the cab roof out of the way when loading. A driver's "uniform" then was either a boiler suit or bib and brace overalls, cap, and an old sports jacket. The chap in the long overcoat with his hands in the pockets is probably the tally clerk or dockers' foreman. The Steers belonged to Coulton of Leyland, and a modern day Health and Safety Inspector would not allow men to stand under the bales on the crane.

Another side view of a Steer showing the weight distribution possible with this axle configuration. The Standard Pulverised Fuel Company was attached to Grimethorpe Colliery, near Barnsley, and the Leyland was carrying coal that had been crushed and ground into a fine powder. Pulverite was delivered as fuel to boilers in smaller establishments that did not have the capability for crushing coal. Larger coal burners such as power stations have their own pulverising mills.

This tipper loaded with crushed stone was probably taken by chance by the official Leyland photographer in Preston town centre. It was heading for Kierby & Perry's ready mixed concrete plant at Bamber Bridge and was one of M. Woodhouse's lorries from Lancaster. It was not a true Steer, but originally a 12.B Beaver that had been converted in-house, re-cabbed, and re-bodied by Grisdale of Churchtown.

The 1965 C&U Regs sounded the death knell for rigid twin steering six wheelers, but by a further quirk of the new regulations this configuration was now ideal for maximum weight articulated lorries. In particular bulk liquids hauliers were quick to place such artics into service. Tilt-cabbed Steers were for 32 tons gross weight and had the Pneumo-cyclic semi-automatic gearbox fitted as standard. They also had lightweight front axles with smaller tyres, easily apparent in this photo. Leyland built very few of these lorries, its sister company AEC supplied both companies' customers who wanted this type with its Mammoth Minor.

A few Steer tractive units were exported and this one was adapted for service in New Zealand. It has a longer wheelbase than its British counterparts and standardised tyre sizes all round. It also had a six speed and deep crawler manual gearbox (10 ratios) in place of a 'Two-Pedal' semi-automatic transmission. In service with Caltex it was photographed at Te Kuiti (North Island) when about 7 years old in 1976. (Photo: Rufus Carr)

This Super Beaver with sleeper cab was used for transporting wine in bulk in Argentina and the picture dates from 1968. In addition to the tanks for the wine it also has racks for carrying crates of bottles. Super Beavers were extremely durable and highly thought of in countries where they were used.

Small numbers of Super Beavers were sold in Great Britian through the years, but to see one in brewery service was highly unusual. Breweries did favour normal control lorries in many instances to enable three men to be carried in the cab and Newcastle Breweries claimed that this lorry was better suited to some narrow lanes in northeastern England. The Wheat Sheaf Hotel is located on the A68 Near Wolsingham, County Durham.

A more typical Super Beaver configuration, as a tractive unit in service in Holland. The lorry has a locally built cab and purpose made radiator muff. The large capacity tank trailer is reminiscent of contemporary Scammell designs.

Super Beaver tractive units sold at home were normally used for abnormal load movements and heavy haulage. Photographed in Lancashire on the A6 heading south from Spade Adam in Cumbria, this Super Beaver was transporting a Blue Streak Missle to the English Electric factory, (now British Aerospace), at Hatfield, Hertfordshire. This was the company's own lorry. Also of interest, the Lancashire Police Force's Ford 'Z' car, which is totally anonymous apart from its beacon, when compared with modern equivalents.

The Leyland Beaver Auto-Tractor was a modified 12.B Beaver for the timber extraction industry. The Auto-Mower Engineering Company Ltd. was licensed to adapt the model by fitting ground anchors, a winch, and crane. For on-road use these specialist lorries would normally tow a round timber pole trailer.

There are always exceptions to every rule and Rollon Transport used this Super Beaver with bulk tipping trailer. It entered service in 1965. This company was part of the Speedway Group, which also included Oliver Hart of Coppull. Traffic consisted of coal and aggregates haulage, and this group had highly individual ideas about lorry design, often modifying vehicles substantially for its own needs.

By most other countries' standards British lorries were conservatively rated in respect of gross weights. Beaver 14.B tractive units were legally limited to 24 tons gross at home, but were designed for more, and in Australia most States allowed higher weights than Great Britain. However, individual axle loadings had to be considered, but the addition of a light tag axle permitted greater loads to be carried.

The location of the third axle on the tractor was varied, and this mid-chassis pusher axle was fitted to a 14.Beaver. It was seen heading south on the original Hume Highway at "The Razorback", south of Camden, near Sydney. Probably heading for Melbourne, it was run by an owner-driver on contract to TNT. This photo was taken in the early seventies when the Leyland was about 15 years old, and a bit long in the tooth to be running interstate routes. (Photo: Phillip Geer)

A type of Beaver never seen in England. Based on Power-Plus models, these modifications were carried out with the approval of, and by, Leyland Motors' Australian company. Some states had an overall articulated vehicle length limit of 45 feet, and "Bridge Formula" regulations relating to axle spacings. The retracted front axle and forward step entry of the long door L.A.D. cab presented a problem in relation to overall length and individual axle loadings. The solution was to fit a short door L.A.D. cab sitting directly over the front axle. With a third axle on the tractor, this is a tag version; a gross vehicle weight of 30.5 tons was permitted. (Photo: Peter Mendoza Studios, from Tony Petch).

Originally supplied to Liquid Cartage of Melbourne, this is another "Comet" cabbed Beaver, and it has a mid chassis pusher axle fitted. It was photographed in Sydney after the large Brambles organisation had taken over Liquid Cartage to build its own bulk haulage division. (Photo: Phillip Geer, from Tony Petch)

SECTION 6

Leyland Beaver in Retrospect

Very few lorries sharing a common model name spanned a period such as that of Leyland's Beavers. Apart from the second world war when production was suspended, they were manufactured for some 45 years. Those earliest Beavers were arguably the first advanced lorries. But such was the eventual pace of change coupled with forced advances in design in the late fifties and mid-sixties final versions had little resemblance to their illustrious predecessors. Beavers bore witness to the birth of long distance road haulage in this country and remained part of the scene to partake in European journeys when they became commonplace many years later.

Nowadays, perhaps it can be difficult to comprehend how different conditions were for transport in the middle decades of last century. Even without implications of legislation the road network itself was barely adequate and these factors influenced aspects of vehicle development, or lack of it. When the Beaver versions available in the thirties are studied one can only be amazed that Leyland Motors offered so many variants. The choice of either petrol or diesel engines is understandable given the suspicions of new technology harboured by sceptical operators. Once the economics and reliability of diesel propulsion became obvious the fate of petrol as fuel for heavy commercials was sealed. And in the thirties Leyland's diesel designs were at least equal to those of some of its competitors and superior to many others. Not only did these early Beavers have proven engines they were robustly constructed in all other respects. They set standards for the reputation of the model name which would be difficult to better. That Leyland Motors were able to do so after the second world war speaks volumes for the company.

It is a sad fact that in wartime despite the unpleasantness and tragic waste wreaked by conflict, technology does advance rapidly. In Leyland Motors' case one positive benefit to emerge during the early forties was the '600' series engine. This became a magnificent power unit and together with its 0.680

derivative became the strong heart of post-war Beavers. From 1946, for the next 14 years, various Beaver models became the pre-eminent four wheeler prime movers. To operate these long wheelbase Beavers as solo machines was really a waste. They were too good for that and needed a trailer to fulfil their true potential. In such a role they were unbeatable and capable of very high mileages over many years of work. As articulation grew in popularity Beaver fifth wheel tractive units also became prevalent and equally successful.

The introduction of Power-Plus Beavers heralded a range of lorries for the dawning Motorway age. Different operating conditions forced buyers to re-appraise their own needs and the growth in articulated unit usage was hastened. Power-Plus Beavers were new in very nearly every aspect of design. Compared with their immediate predecessors some engine reliability was lost in the quest for extra power. For example P.680 units were prone to cylinder head gasket failures if continuously driven to their limits. One other criticism of Power-Plus lorries was the L.A.D. styled cab. While it was pleasing to look at and projected a modern image, its interior was rather cramped and lacked space. Some purchasers grumbled about paying premium prices for top quality chassis and drivelines only to obtain cabs similar to those used by one maker of much cheaper lorries.

Leyland Motors recognised these shortcomings of L.A.D. cabs and used them on Beavers for a relatively short period. When Freightline Beavers were announced in 1964 the Ergomatic tilting cab gave drivers a de-luxe environment of comfort and quietness superior to structures available from any other British, non-Leyland Group maker. Freightline Beaver chassis and drivelines were based on Power-Plus models, although a Mk.II version of the Leyland P.680 engine rectified some of the reliability concerns. Legislation enacted in 1965 and 1968 resulted in an explosion in demand for artics and the vast majority of Freightline Beavers were built to tractive unit specifications.

The most exciting Freightline Beavers were the Two-Pedal variants. Advanced and innovative lorries for their time, inevitably there were teething problems. None were insurmountable but the parent company was riven by mounting financial problems and industrial relations disputes in its other divisions. This adversely affected the heavy vehicles side of the business and 'Two-Pedal' systems were abandoned. Recollections from drivers of these Beavers confirm they were easy and lovely lorries to drive. While they were "one driver" vehicles nevertheless they gave excellent performance if handled correctly. What a great pity the concept was not persevered with and perfected.

Leyland Beavers and Super Beavers were exported in quantity for many years. Their attributes of durability and reliability made them popular in countries like Australia and South Africa. They were also found in numerous other regions where Leyland Motors had representation. Early Beavers were instrumental in establishing Leyland's reputation for top quality. Subsequent models enhanced the firm's esteem in its halcyon years.

As has been stated previously the Leyland 0.600 engine was one of the great power units of its era and was versatile in many and varied applications. It ranks alongside contemporary engines such as AEC's 9.6 litre, and Gardner's LW series as a truly world class design. Leyland engine expertise was much sought after by some European and Scandinavian commercial vehicle makers. In particular Beaver 0.600 and 0.680 engines helped to establish at least two lorry builders into huge organisations today. DAF of Holland arose from being a builder of trailers to a fledgling lorry maker, and then by using Leyland engines became a well respected name. After many troubled years for British Leyland, DAF rescued the once proud old firm. But what despair was caused by their decision in the year 2000 to abandon the Leyland name. A typical marketing decision made by hard faced executives who care nothing for tradition or heritage. These people know the price of everything but the value of nothing and they forget the lessons of history at their peril.

Our industrial legacy can justly claim to having a highly respected commercial vehicles industry for most of the twentieth century. Leyland Motors was one of the noblest firms involved. This famous name and its products were known and revered throughout the world. Leyland's attributes of practical design coupled with solid engineering made robust, durable lorries which gave operators reliability, longevity, and economic service. For over 40 years Leyland Beavers epitomised these principles. Especially when pulling a drawbar trailer they can be regarded as definitive British lorries of their age.

APPENDIX A
Leyland Beaver Chassis Details

Type & Description	Control Layout	Gross Weight	Engine	Year
BEAVER TC	Normal	$3^{1}/_{2}$ Tons load	4 cyl Petrol	1928
BEAVER 4 TC 10, 11, 12 BEAVER 4 TSC 11, 12, 13	Normal Forward	12 Tons	4 cyl Petrol or Diesel	1933
BEAVER 6 TC 9, 9A, 13, 14 BEAVER 6 TSC 8, 9, 10	Normal Forward	12 Tons Solo 19 Tons as a Prime Mover	6 cyl Petrol or Diesel	1933
BEAVER4 TC 10A, 11A, 12A BEAVER 4 TSC 11A, 12A, 13A	Normal Forward	12 Tons Solo 14½ Tons as a Prime Mover	4 cyl Petrol or Diesel	1937
BEAVER 6 TRACTOR TC 14A	Normal	19 Tons	6 cyl Petrol or Diesel	1937
"INTERIM" BEAVER	Forward	12 Tons Solo 19 Tons as a Prime Mover	6 cyl Diesel 7.4 Litre	1945
BEAVER 12.B/1 Haulage BEAVER 12.B/3 Tipper BEAVER 12.B/7 Tractive Unit	Forward	12 Tons Solo 19 Tons as a Prime Mover	0.600	1946
AIR BRAKED MODELS BEAVER 12.B/8 Haulage BEAVER 12.B/9 Tipper BEAVER 12.B/10 Tractive Unit N.B. 'A' Suffix Would Denote Modernised Cab	Forward	12 Tons Solo 19 Tons as a Prime Mover	0.600 or 0.680	1953
BEAVER 14.B/8 Haulage BEAVER 14.B/9 Tipper BEAVER 14.B/10 Tractive Unit	Forward	14 Tons Solo 24 Tons as a Prime Mover	0.600 or 0.680	1955
EXPORT BEAVERS 14.B1E Haulage 14.B3E Tipper 14.B7E Tractive Unit	Forward	14 Tons Solo 24 Tons as a Prime Mover	0.600 or 0.680	1957
POWER-PLUS BEAVERS (Home Market) 14B.11R, 12R, 13R Haulage 14B.14R Tipper 14B.17R Tractor	Forward	14 Tons Solo 24 Tons as a Prime Mover	P.600 or P.680	1960

(N.B. 'L' SUFFIX = EXPORT, LEFT HAND CONTROL)

FREIGHTLINE BEAVERS (Home)

16BT.1R, 2R Haulage		16 Tons	P.600	
16BT.3R, 4R Tipper	Forward	16 Tons	or	1965
14BT.17R, 18R, Tractor		30 Tons	P.680	

FREIGHTLINE BEAVERS (Export)

16BT.1L, 2L Haulage		16 Tons		
16BT.3L, 4L Tipper	Forward	16 Tons	P.680	1965
14BT.17L, 18L Tractor		30 Tons		

TWO-PEDAL BEAVERS

14BT.28R Tractor		30 Tons	P.680	1966
BE.TR.1PR Tractor	Forward	30/32 Tons	P.680	1967
BV68.32PTR Tractor		32 Tons	P.680	1968
BV69.32PTR Tractor		32 Tons	690	1968

FREIGHTLINE BEAVERS (Home)

16BT.1CR, 2CR Haulage		16 Tons	P.680 or P.600	
16BT.4CR Tipper	Forward	16 Tons	P.680 or P.600	
14BT.17BR Tractor		30 Tons	P.680	1968
14BT.18 BR Tractor		32 Tons	P.680	

FREIGHTLINE BEAVERS (Export)

16BT.1AL, 2AL Haulage		16 Tons	P.680 or P.600	
16BT.3AL, 4AL Tipper	Forward	16 Tons	P.680 or P.600	1968
14BT.17AL Tractor		30/32 Tons	P.680	

RATIONALISED BEAVERS

16BT.1DR Haulage (Home)		16 Tons	P.680	
16BT.1BL Haulage (Export)	Forward	16 Tons	P.680	
BV68.32TR Tractor		32 Tons	P.680	1970
BV69.32TR Tractor		32 Tons	690	

STEERS

STEER TEC 1 Haulage	Forward	15½ Tons Solo or	6 cyl Petrol or	1937
STEER TEC 2 Haulage		22 Tons Drawbar	6 cyl Diesel	1937
STEER 15.S/1 Haulage	Forward	15½ Tons Solo or 22 Tons Drawbar	0.600	1947
STEER 16.S3 Haulage (Home)	Forward	16½ Tons Solo or 24 Tons Drawbar	0.600 or 0.680	1955
STEER 16.S3E Haulage (Export)				
FREIGHTLINE STEER Tractor	Forward	32 Tons	P.680	1966

SUPER BEAVERS

EB.3ARE Haulage		12 to 16 Tons Solo	0.600 From	1947
EB.3AL Haulage		or 24 to 32 Tons	Either 0.600	
EB.7ARE Tipper	Normal	as a Prime Mover	or 0.680 From	1951
EB.9AL Tipper		Gross Weight by	0.680 From	1955
EB.11ARE Tractor		Tyre Size Fitted	P.680 From	1960
EB.12AL Tractor		and Local Regs.		

NOTES: 1. Steer 15.S/1 Model had optional 0.680 engine available from 1951
2. Freightline Rigid Beavers for drawbar duties, gross train weight of 30/32 tons.

APPENDIX B

Beaver Engine Details

Power outputs have been corrected to a common data base which is the British Standards AU:141 1967 rating. This is to provide meaningful comparisons between pre-war and later engines. BS conditions were standardised at 60 degrees Farenheit and 29.92 inches of mercury atmospheric pressure. Brake horse power is nett power available at the flywheel after deducting that absorbed by engine auxiliaries. For further comparison with modern metric bhp increase these quoted figures by 9%.

TYPE	LAYOUT	CYLs.	IMPERIAL DIMENSIONS			RAC H.P.	METRIC DIMENSIONS			BHP	@RPM	TORQUE Lb.Ft	@RPM	YEAR
			BORE	STROKE	CAP. Cu. Ins.		BORE	STROKE	CAP. cc					
Petrol	OHC	4	4.0"	5½"	276	27.2	101.5mm	139.7mm	4,533	62	2,200	200	1,000	1928
Petrol	OHC	4	4¼"	5½"	312	28.9	107.9mm	139.7mm	5,112	65	2,200	225	1,000	1930
Petrol	OHC	4	4.56"	5½"	360	33.3	115.9mm	139.7mm	5,893	70	2,200	250	1,100	1931
Petrol	OHC	6	4¼"	5½"	468	43.5	107.9mm	139.7mm	7,670	110	2,200	320	1,000	1931
Petrol	OHC	6	4.56"	5½"	540	50.0	115.9mm	139.7mm	8,840	115	2,200	360	1,000	1931
Petrol Pushrod Mk.III		6	4.0"	5½"	415	38.4	101.6mm	139.7mm	6,800	105	2,300	300	1,000	1937
Petrol Pushrod Mk.III		6	4.31"	5½"	482	44.6	109.5mm	139.7mm	7,900	116	2,200	350	1,000	1937
Oil	OHC	4	4½"	5½"	350	32.4	114.3mm	139.7mm	5,731	68	1,800	205	1,300	1933
Oil	OHC	4	4½"	5½"	350	32.4	114.3mm	139.7mm	5,731	71	1,900	210	1,300	1934
Oil	OHC	6	4½"	5½"	525	48.6	114.3mm	139.7mm	8,600	93	1,900	305	1,300	1934
Oil	OHC	6	4½"	5½"	525	48.6	114.3mm	139.7mm	8,600	106	1,900	330	1,200	1938
Diesel Pushrod 7.4		6	4.37"	5.0"	451	N/A	110.9mm	127.0mm	7,390	100	1,800	315	1,100	1945
Diesel Pushrod 0.600		6	4.8"	5½"	597	N/A	122.0mm	139.7mm	9,800	125	1,800	410	900	1946
Diesel Pushrod 0.680		6	5.0"	5¾"	677	N/A	127.0mm	146.0mm	11,100	150	2,000	450	1,100	1951
Diesel Pushrod P.600		6	4.8"	5½"	597	N/A	122.0mm	139.7mm	9,800	140	1,700	438	1,200	1960
Diesel Pushrod P.680		6	5.0"	5¾"	677	N/A	127.0mm	146.0mm	11,100	200	2,200	548	1,200	1960
Diesel Pushrod 690		6	5.0"	5¾"	677	N/A	127.0mm	146.0mm	11,100	240	2,200	650	1,400	1968